27305-14

AF217990

Handling and Placing Concrete

OVERVIEW

On large commercial and industrial projects, carpenters are involved to some extent in the handling and placing of concrete. This module provides information on the equipment and methods used to place and finish concrete, as well as the various types of joints used to control cracking.

Module Nine

Trainees with successful module completions may be eligible for credentialing through NCCER's National Registry. To learn more, go to **www.nccer.org** or contact us at **1.888.622.3720**. Our website has information on the latest product releases and training, as well as online versions of our *Cornerstone* magazine and Pearson's product catalog.

Your feedback is welcome. You may email your comments to **curriculum@nccer.org,** send general comments and inquiries to **info@nccer.org**, or fill in the User Update form at the back of this module.

This information is general in nature and intended for training purposes only. Actual performance of activities described in this manual requires compliance with all applicable operating, service, maintenance, and safety procedures under the direction of qualified personnel. References in this manual to patented or proprietary devices do not constitute a recommendation of their use.

Objectives

When you have completed this module, you will be able to do the following:

1. List the safety precautions for handling, placing, and finishing concrete.
   a. List the rules for the care and safe use of hand tools when handling and placing concrete.
   b. List the rules for the care and safe use of power tools when handling and placing concrete.
   c. Explain how to prevent cement dermatitis.
2. Identify the methods of moving and handling concrete.
   a. Identify off-site equipment for mixing and conveying concrete.
   b. Identify on-site equipment for mixing and conveying concrete.
   c. Explain how to use hand and power tools for mixing and conveying concrete.
3. Explain the proper methods for placing and consolidating concrete into forms.
   a. Explain the proper method for placing concrete into forms.
   b. Explain the proper method for consolidating concrete.
4. Describe the proper methods for finishing and curing concrete.
   a. Explain the proper method for screeding concrete.
   b. Explain the proper method for leveling concrete.
   c. Explain the proper method for finishing concrete.
   d. Describe how to properly cure concrete.
   e. Describe the use of joint sealants.
   f. Identify the tools used to rub and patch concrete.
5. Identify the different kinds of joints in concrete structures.
   a. Identify construction joints.
   b. Identify isolation joints.
   c. Identify control joints.
   d. Identify decorative joints.

Performance Tasks

Under the supervision of your instructor, you should be able to do the following:

1. Properly place and consolidate concrete in selected concrete forms.
2. Use a screed to strike off and level a concrete surface.
3. Use a bull float and/or darby to level and smooth a concrete surface.
4. Use an edger to form a radius at the edges of a concrete pad, slab, etc.
5. Use a hand float and finishing trowel to level high spots, remove imperfections, and smooth a concrete surface.
6. Use a jointer to make control joints in a concrete surface.

Trade Terms

Articulating	Elastomeric	Segregation
Bleed	Grout	Set
Bulkhead	Laitance	Shrink mixing
Consolidating	Lift	Spalling

Industry Recognized Credentials

If you're training through an NCCER-accredited sponsor you may be eligible for credentials from NCCER's Registry. The ID number for this module is 27305-14. Note that this module may have been used in other NCCER curricula and may apply to other level completions. Contact NCCER's Registry at 888.622.3720 or go to **www.nccer.org** for more information.

Code Note

Codes vary among jurisdictions. Because of the variations in code, consult the applicable code whenever regulations are in question. Referring to an incorrect set of codes can cause as much trouble as failing to reference codes altogether. Obtain, review, and familiarize yourself with your local adopted code.

Contents

Topics to be presented in this module include:

Figures and Tables

1.0.0 SAFETY PRECAUTIONS FOR HANDLING, PLACING, AND FINISHING CONCRETE

Objective

List the safety precautions for handling, placing, and finishing concrete.
  a. List the rules for the care and safe use of hand tools when handling and placing concrete.
  b. List the rules for the care and safe use of power tools when handling and placing concrete.
  c. Explain how to prevent cement dermatitis.

Using concrete tools for their intended purpose, using them safely, and properly caring for tools is the mark of a professional. Always follow Occupational Safety and Health Administration (OSHA) regulations and the manufacturer's safety precautions. Protect yourself and co-workers by following the rules for proper care and use of hand and power tools. Skin exposure to concrete or mortar can cause cement dermatitis. Protect exposed skin by wearing the proper protective clothing when working with concrete.

Concrete is a heavy building material. Care must be taken to avoid strains of the back, arms, and legs. Prior to beginning work with concrete, perform proper stretching exercises to minimize the potential of strains or sprains.

1.1.0 Hand-Tool Safety

Some rules for the care and safe use of hand tools are as follows:

- Use the correct tool for the job.
- Keep tools clean; be sure to wash concrete from surfaces immediately after use.
- Maintain tools properly.
- Sharpen tools, when appropriate.
- Inspect tools frequently to make sure they are in good condition.
- Repair or replace broken or damaged tools promptly.
- Dispose of broken tools that cannot be repaired.
- Do not use broken tools.
- Do not throw tools.

- Protect the cutting edge of tools when carrying and storing them.
- Store tools properly when not in use. Lightly oil tools before storing. Store tools in a dry place.
- Do not carry tools in your pocket.
- Do not place tools on surfaces where they can roll off.
- Wear appropriate personal protective equipment, including eye, ear, and respiratory protection.
- Stay alert when using tools. Keep fingers away from cutting edges. Work away from your body when using cutting tools.
- Be sure everyone is clear before you swing a sledgehammer.
- Use tools with insulated or wood handles when working around electrical equipment.

1.2.0 Power-Tool Safety

Some rules for the care and safe use of power tools are as follows:

- Do not attempt to operate any power tool without being trained and certified on that particular tool.
- Always wear appropriate personal protective equipment and protective clothing. For example, wear safety glasses and tight-fitting clothing that cannot become caught in the moving parts of the tool. Roll up long sleeves, tuck in your shirt tail, and tie back long hair.
- Do not leave a power tool running unattended.
- Assume a safe and comfortable position before starting a power tool.
- Do not distract others or let anyone distract you while operating a power tool.
- Be sure that any corded power tool is properly grounded before using it.
- Be sure that power is not connected before performing maintenance or changing accessories.
- Do not use dull or broken accessories.
- Use a power tool only for its intended purpose.
- Do not use a power tool with guards or safety devices removed.
- Use an extension cord of sufficient length to service the particular electric tool you are using.
- Do not operate a corded power tool if your hands or feet are wet.
- Become familiar with the correct operation and adjustments of a power tool before attempting to use it.
- Keep a firm grip on the power-float or trowel-machine handle while operating the machine.
- Ensure there is proper ventilation before operating gasoline-powered equipment indoors.

- Keep a fire extinguisher nearby when filling and operating gasoline-powered equipment.
- Keep hands and feet away from cutting tools such as concrete saws, grinders, and trowel-machine blades.
- Change trowel-machine blades when they become ragged.
- Store tools properly when not in use.

1.3.0 Cement Dermatitis

Persons working with concrete need to take precautions to prevent skin irritation as a result of skin coming into contact with concrete or mortar. When wet, the lime contained in cement causes it to heat and burn the skin. Prolonged exposure and inhalation could result in sores appearing on the skin and in the lining of the mouth and nose. This condition is called cement poisoning or cement dermatitis. Factors that contribute to its development include excessive sweating, failure to observe precautions, and pre-existing dermatitis or allergy. Medical care should be sought by anyone who develops dermatitis. In general, proper personal protective equipment (PPE) will reduce, but not eliminate, most exposure incidents. Always refer to the manufacturer's safety data sheet (SDS) for information on the proper PPE to be used. Some actions that help reduce the risk of cement dermatitis include the following:

- Use of the proper clothing such as coveralls and long-sleeved shirts
- Use of rubber gloves and boots
- Use of goggles or face mask
- Use of a respirator or other breathing device
- Frequent washing with a lanolin-based soap or lotion
- Bathing after each shift

1.0.0 Section Review

1. Tools should *not* be carried _____.

   a. over your shoulder
   b. in your pocket
   c. while climbing stairs
   d. in a wheelbarrow

2. Do *not* operate a corded power tool _____.

   a. if your hands are wet
   b. plugged into an extension cord
   c. without being certified by the manufacturer
   d. that has not been used for an extended period

3. Wet concrete can cause a skin irritation called cement _____.

   a. rash
   b. dermatosis
   c. dermatitis
   d. burn

NCCER – *Carpentry Level Three* 27305-14

SECTION TWO

2.0.0 MOVING AND HANDLING CONCRETE

Objective

Identify the methods of moving and handling concrete.
  a. Identify off-site equipment for mixing and conveying concrete.
  b. Identify on-site equipment for mixing and conveying concrete.
  c. Explain how to use hand and power tools for mixing and conveying concrete.

Trade Terms

Articulating: Having a hinged or pivoting connection.

Consolidating: Working freshly placed concrete so that each layer is compacted with the layer below and voids caused by water or air pockets are eliminated.

Grout: A thin mortar.

Segregation: The separation of sand-cement ingredients from the gravel due to the improper placement of concrete.

Shrink mixing: Practice of mixing concrete at a batch plant to the point where the slump meter indicates that the desired slump is predictable, then finishing mixing the concrete on the way to the job site.

Advance planning is necessary in order to avoid costly delays or problems in placing or finishing concrete. To achieve the best productivity and job quality, the work must be planned to make the best use of the material, personnel, and equipment. All preparatory steps must be completed; the proper tools and personnel must be available; weather conditions must be favorable; and the concrete must be ordered and available when needed.

The basic tasks involved with the placing and finishing of concrete are:

- Moving and handling concrete
- Placing concrete in forms
- Consolidating concrete
- Finishing

The transportation and handling of concrete involves the use of both off-site and on-site equipment. Off-site equipment is used to mix and transport the concrete from the concrete plant to the job site. On-site equipment is used at the job site to mix, move, and place the concrete.

2.1.0 Mixing and Conveying Concrete Using Off-Site Equipment

Typically, commercial concrete used for construction is produced to specifications off-site at a batch plant or central mix plant. Components at both types of plants typically include compartmented storage bins of the different grades of aggregate, storage bins for each type of cement, weight batchers for proportioning the cement and aggregate, a means of batching water, and some means for controlling the batching process. The arrangement and size of the components is determined by the design and type of plant. Detailed information about the composition of concrete and the different types of concrete mixtures is covered in the *Properties of Concrete* module.

Batch plants (*Figure 1*) are dry-mix plants, meaning that they only proportion or batch the ingredients, but do not mix them. The actual mixing of the concrete is performed in the trucks, called truck mixers, which deliver the concrete to the job site. Because the concrete is mixed in the truck, the mixing can be controlled so that the concrete can be transported for short or long distances before it must be discharged. Truck mixers used to transport concrete can be a rear-discharge or front-discharge type (*Figure 2*). Front-discharge mixers are popular because they are highly maneuverable and allow the driver to distribute the concrete by moving the chute via a hydraulic system without leaving the truck cab.

Central mix plants are wet-mix plants, meaning they produce fully mixed plastic concrete. Once mixed, the concrete is delivered to the job

27305-14_F01.EPS

Figure 1 Concrete batch plant.

27305-14_F02.EPS

Figure 2 Front-discharge truck mixer.

27305-14_F03.EPS

Figure 3 Mobile concrete batch plant.

site via truck mixers. While in transit, the plastic cement is prevented from setting up prematurely via agitation by the truck's mixing mechanism.

With plant-mixed concrete, it is a common practice to mix the concrete to the point where the plant slump meter indicates that the desired slump is predictable, then finish mixing the concrete on the way to the job site. This procedure is called shrink mixing. Because the concrete is plastic when it leaves the plant, the distance that it can be transported before it must be discharged is normally less than that allowed for truck-mixed concrete.

2.2.0 Mixing and Conveying Concrete Using On-Site Equipment

When the job site is in a remote area where concrete delivery is not readily available, or where sufficient quantities of concrete required for the project justify their cost, mobile concrete batching and mixing equipment located at the job site are often used. Some other reasons for using on-site mobile concrete production equipment include: the elimination of waiting times for mixer delivery trucks to arrive, the ability to customize the concrete mixture on demand for different on-site pour requirements, and the ability to limit the amount of concrete produced to an as-needed basis. Mobile concrete batching plants and volumetric mixers are two types of mobile equipment commonly used to produce concrete at the job site. Mobile concrete batch plants like the one shown in *Figure 3* are mounted on one or more trailers for transportation to and from the job site. Mobile concrete batch plants all have top-loading bins for holding aggregates and cement, and water tanks for holding water. Depending on the manufacturer, model, and design, mobile concrete batch plants have varying concrete production capacities.

Volumetric mixers are truck-mounted mixers (*Figure 4*) that have water tanks and top-loading

bins for holding coarse aggregate, fine aggregate, and cement. Correctly proportioned dry coarse aggregate, fine aggregate, and cement supplied from the bins simultaneously drop onto and are carried by conveyors into the charging end of the mixer at the rear of the unit. There, a predetermined metered flow of water also enters the mixer. Via the action of a mixer assembly, the dry ingredients and water are rapidly and thoroughly mixed to produce a continuous discharge of concrete. The mixing and discharging of concrete can be stopped at any time and started again as determined by the vehicle operator. The capability to start and stop mixing and discharging of concrete allows the production of concrete to be controlled to the needs of the placing and finishing crews and other job requirements. It also allows the truck to move to different concrete placing locations within the job site.

Concrete is moved around the job site and placed using a wide variety of equipment. The specific equipment used depends on the type of construction and the topography of the job site. Typically, the equipment used includes buckets handled by cranes, concrete pumps, chutes, and buggies. Concrete conveying equipment is summarized in *Table 1*.

27305-14_F04.EPS

Figure 4 Volumetric mixer truck.

NCCER – *Carpentry Level Three* 27305-14

Table 1 Methods of Conveying Concrete (1 of 2)

Equipment	Type and Range of Work for Which Equipment Is Best Suited	Advantages	Points to Watch For
Truck agitator	Used to transport concrete for all uses in pavements, structures, and buildings. Haul distances must allow discharge of concrete within 1½ hours, but this limit may be waived under certain circumstances.	Usually operate from central mixing plants where quality concrete is produced under controlled conditions. Discharge from agitators is well-controlled, with uniformity of concrete on discharge.	Timing of deliveries to suit job organization. Concrete crew and equipment must be ready on site to handle concrete. Large batches.
Truck mixer	Used to mix and transport concrete to job site over short and long hauls. Hauls can be any distance.	No central mixing plant is needed, only a batching plant, since concrete is completely mixed in the truck mixer. Discharge is the same as for truck agitator.	Control of concrete quality is not as good as with central mixing. Slump tests of concrete consistency are needed on discharge. Careful preparations are needed for receiving the concrete.
Nonagitating truck	Used to transport concrete on short hauls.	Capital cost of nonagitating equipment is lower than that of truck agitators or mixers.	Concrete slump should be limited. Possibility of segregation. Height is needed for high lift of truck body upon discharge.
Mobile continuous mixer	Used for continuous production of concrete at job site.	Combination of materials transporter and mobile mixing system for quick, precise proportioning of specified concrete. One-person operation.	Trouble-free operation requires good preventive maintenance program on equipment. Materials must be identical to those in original mix-design proportioning.
Crane	Used for work above and below ground level.	Can handle concrete reinforcing steel formwork and sundry items in high-rise concrete-framed buildings.	Has only one hook. Careful scheduling between trades and operations is needed to keep it busy.
Buckets	Used on cranes and cableways for construction of buildings and dams. For conveying concrete directly from central discharge point to form work or to secondary discharge point.	Enable full versatility of cranes and cableways to be exploited. Clean discharge. Wide range of capacities.	Select bucket capacity to conform with size of the concrete batch and capacity of the placing equipment. Discharge should be controllable.
Barrows and buggies	For short, flat hauls on all types of on-site concrete construction, especially where accessibility to work area is restricted.	Very versatile and therefore ideal inside and on job sites where placing conditions are constantly changing.	Slow and labor intensive.
Chutes	For conveying concrete to lower level (usually below ground level) on all types of concrete construction.	Low cost and easy to maneuver. No power required. Gravity does most of the work.	Slopes range between 1 to 2 and 1 to 3 and chutes must be adequately supported in all positions. Arrange for discharge at end (downpipe) to prevent segregation.
Belt conveyors	For conveying concrete horizontally or vertically. Usually used between main discharge point and secondary discharge point. Not suitable for conveying concrete directly into formwork.	Belt conveyors have adjustable-reach traveling diverter, and variable speed both forward and reverse. Can place large volumes of concrete quickly when access is limited.	End-discharge arrangements needed to prevent segregation. Leave no mortar on return belt. In adverse weather (hot or windy) long reaches of belt need cover.

Table 1 Methods of Conveying Concrete (2 of 2)

Equipment	Type and Range of Work for Which Equipment Is Best Suited	Advantages	Points to Watch For
Pneumatic guns	Used where concrete is to be placed in difficult locations and where thin sections and large areas are needed.	Ideal for placing concrete in free-form shapes, for repairing and strengthening buildings, and for protective coatings and thin linings.	Quality of work depends on the skill of those using the equipment. Only experienced nozzle operators should be employed.
Concrete pumps	Used to convey concrete directly from central discharge point to formwork or secondary discharge point.	Pipelines take up little space and can be readily extended. They deliver concrete in a continuous stream. Mobile boom pump can move concrete both vertically and horizontally.	Constant supply of fresh, plastic concrete is needed with average consistency and without any tendency to segregate. Care must be taken in operating pipeline to ensure an even flow and to clean out at conclusion of each operation. Pumping vertically around bends and through flexible hose will considerably reduce the maximum pumping distance.
Drop chutes	Used for placing concrete in vertical forms of all kinds. Some chutes are in one piece, while others are assembled from a number of loosely connected segments.	Direct concrete into formwork and carry it down to the bottom of forms without segregation. Their use avoids spillage of grout and concrete on form sides, which is harmful when off-the-form surfaces are specified. They also prevent segregation of coarse particles.	Should have sufficiently large, splayed top openings into which concrete can be discharged without spillage. The cross section of a drop chute should be chosen to permit inserting into the formwork without interfering with steel reinforcing.
Tremies (elephant trunks)	Used with drop chute to extend its reach so it can place concrete into narrow wall forms, or at the bottom of a deep form such as a high column form.	Flexibility allows it to place concrete lower in a form to minimize segregation.	Be careful not to let rebar damage elephant trunk, as it is made of flexible material.
Screw spreaders	Used for spreading concrete over flat areas, as in pavements.	A batch of concrete discharged from bucket or truck can be spread quickly over a wide area to a uniform depth. The spread concrete has good uniformity of compaction before vibration is used for final compaction.	Screws are usually used as part of a paving train. They should be used for spreading before vibration is applied.

Using improper placement methods or equipment when placing concrete can result in **segregation** of the concrete materials. Segregation is the tendency for the coarse aggregates to separate from the sand-cement mortar. This results in part of the batch having too little coarse aggregate and the remainder having too much. The former is likely to shrink and crack and have poor wear resistance. The latter is too harsh for full consolidation and finishing. Segregation can usually be avoided by following these placing procedures:

- The drop of the concrete should always be vertical.
- Drop chutes should be used to prevent concrete from striking the reinforcement steel or the side of the form above the level of placement.
- The free-fall distance of the concrete should not exceed 4 feet.

2.2.1 Cranes and Buckets

Cranes equipped with concrete buckets are widely used to transport concrete to levels above or below ground level. On very large projects, this is often done using a tower-type crane similar to the one shown in *Figure 5*. Tower cranes can be rotated 360 degrees, allowing them to reach all areas of a building being constructed. Tower cranes are erected at the job site. When the building is complete, they are dismantled and taken away. If the project being built is not large enough or tall enough to warrant the erection of a tower crane, then portable cranes are brought to the job site to handle concrete as well as other building materials.

A concrete bucket is used to transport concrete from one location to another when using either type of crane. *Figure 6* shows a conventional bucket. Buckets can be either round or square and are made with capacities ranging from $\frac{1}{3}$ to 12 cubic yards. There are many bucket designs; they differ in the slopes to their sides and the sizes of the gate openings. Discharge from the bucket may be either by hand-operated gates for smaller buckets or air-actuated gates for larger buckets.

27305-14_F05.EPS

Figure 5 Tower cranes.

WARNING! Concrete weighs approximately 4,150 pounds per cubic yard. The combined weight of the bucket and its contents must never exceed the safe capacity of the crane and its associated rigging.

2.2.2 Concrete Pumps

Today mobile boom-type concrete pumps (*Figure 7*) are widely used both on commercial and residential jobs to place concrete. They are used to pump concrete from ready-mix delivery trucks horizontally and/or vertically to placing locations above, below, and at ground level. Typical applications include the placement of large volumes of concrete for high-rise buildings and other commercial and industrial jobs and for large slabs. Because of their reach and their **articulating** boom arms that can rotate 360 degrees, the pump trucks can often be placed at one location on a job site to accomplish the concrete placement for an entire pour. This allows the concrete loads from a series of ready-mix delivery trucks to be discharged directly into the hopper of the boom concrete pump all at one location.

Many sizes and configurations of boom concrete pumps are available. These can range from single-axle truck-mounted pumps to large multi-axle units used for their powerful pumps and long-reach booms. Boom configurations typically range from three to five sections. The typical concrete pump used for high-rise construction can pump over 200 cubic yards of concrete per hour and has the capability to pump the concrete more than 2,500 feet horizontally and over 1,000 feet vertically. Most have lines 5 inches in

27305-14_F06.EPS

Figure 6 Concrete bucket.

27305-14_F07.EPS

Figure 7 Boom-type concrete pump.

diameter or less. Concrete pumps equipped with a rubber-end hose are used to place concrete in wall forms and other areas where projecting vertical reinforcement does not allow the use of drop chutes or tremies. Operation of a concrete pump is performed by an experienced operator. Positioning of the pump boom during boom erection, concrete placement, and boom removal is usually done by the pump operator working with a spotter located at the placement site. The operator and spotter communicate using two-way radios. Working with a spotter is recommended in order to achieve correct boom placement and to avoid boom contact with overhead power lines and other obstructions.

WARNING!

When placing concrete using a concrete pump, dead time must be avoided because concrete can harden in the boom. If the concrete is allowed to harden, the boom may explode or a mass of concrete may be projected from the end of the hose at a high velocity.

Some important factors to take into consideration when using a boom concrete pump include:

- *Overhead power lines* – The American Concrete Pumping Association requires that the boom tip and other sections of the boom be at least 17 feet from power lines. The boom chassis and many of its other components are conductors of electricity. The plastic concrete being pumped can also become a conductor of electricity because of its high water content.
- *Excavations* – Proper placement of a boom-type concrete pump near an excavation or land that falls off steeply is also important. A general rule of thumb states that when pumping concrete near a vertical drop-off of an excavation, the pump should be located a minimum of 1 foot back from the edge of the excavation for every foot in depth of the excavation. For example, if the depth of an excavation is 20 feet, the pump should be located a minimum of 20 feet from the edge of the excavation.
- *Pump-truck stabilization* – Proper-sized load-bearing supports called cribbing must be placed under each of the pump truck's outriggers to stabilize the pump trunk while pumping concrete. The area of cribbing under each outrigger must be big enough so that the pressure placed on each outrigger is less than the loadbearing capability of the soil. One method of determining the proper cribbing size involves laying cribbing on firm spots of ground and positioning the pump truck's outriggers on them. Following this and one at a time, the boom is extended over each outrigger and a check is made to see if the cribbing sinks into the soil. If it does, the boom must be refolded, and a larger area of cribbing placed under the outrigger. Typical cribbing materials used include steel or aluminum plates or layers of 4" × 4" or 4" × 6" timbers.
- *Pump pipeline diameter* – The diameter of the pump pipeline used for an operation is based on the largest size of the aggregate in the concrete that is to be pumped. As a rule of thumb, the pipe size should be at least three times as large as the maximum-size aggregate in the concrete mixture in order to minimize the

Discharging Concrete from a Bucket

To prevent segregation, concrete being discharged from a bucket should not be allowed to drop freely in the air more than 4 feet.

risk of clogging in the pump and pipeline. The larger the diameter of the aggregate in the mixture, the larger the pump and pipe size required to pump the concrete.

- *Pipeline length and layout* – Ideally, the boom pipeline length should be as short as possible. This is because line length directly reflects the pump line pressure. Also, to maintain the least resistance, the pipeline should contain the minimum amount of bends possible.

2.2.3 Chutes

Chutes are a widely used and simple way of transferring concrete to a lower elevation. The best example is the movable semicircular chute used to discharge concrete from the truck mixer into a bucket or other equipment, or directly into the forms. Chutes must be positioned so that they have sufficient slope to allow the concrete to readily move down them by the force of gravity.

2.2.4 Drop Chutes and Tremies

Drop chutes and tremies (elephant trunks) are used to place concrete at lower elevations without causing segregation. See *Figure 8*. Segregation can occur by the concrete hitting the form walls, reinforcing steel, or other obstructions. Drop chutes and/or tremies are typically used when placing concrete into narrow wall forms, or at the bottom of a deep form such as a high column form. They can also be used at the end of a conveyor belt or concrete pump. Drop chutes are typically made of sheet metal or short sections of steel fastened together. Tremies are made of rubber or plastic tubing so that they are flexible and can readily be shortened. The hoppers used to distribute concrete into drop chutes or tremies must be large enough and steep enough to allow for the quick discharge of the concrete without plugging.

2.2.5 Wheelbarrows, Power Buggies, and Carts

Wheelbarrows, concrete carts, and power buggies (*Figure 9*) can also be used to transport concrete from the truck mixer to the placement location.

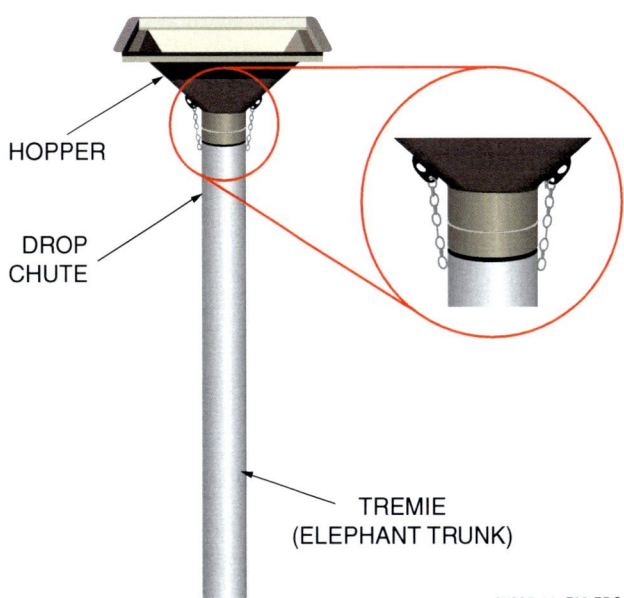

Figure 8 Hopper, drop chute, and elephant trunk.

Typically, the use of these devices is limited to small jobs, on level ground or floor surfaces, and over short distances. Whenever wheelbarrows, concrete carts, or power buggies are used, their travel should be over runways built for that purpose. The runways should be constructed of boards or plywood and made smooth to prevent bumps that tend to cause the concrete to segregate.

2.2.6 Belt Conveyors

Belt conveyors are used to transport concrete horizontally and for short vertical distances. They are typically used to transfer concrete from a truck mixer to a hopper on a concrete pump or drop chute, and into wheelbarrows, concrete carts, or power buggies. They are ideal where the ground conditions do not allow the mixer to get close to the placement location.

2.2.7 Shotcrete

Another method of transferring concrete to a surface is by spraying. When cement mortar is sprayed onto a surface under pneumatic pressure, the product is called shotcrete. Shotcrete usually requires no forms. Shotcrete has a superior bond-

Boom-Type Concrete-Pump Accidents

OSHA states that more concrete-pump operators die from electrocution than any other job-related cause. More than 50 percent of accidents involving boom-type concrete pumps and power lines happen when the boom is being folded, unfolded, or removed.

WHEELBARROW

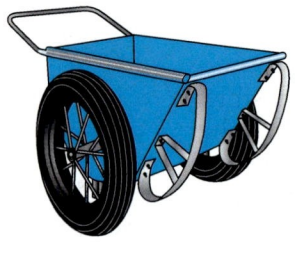
CART

POWER BUGGY

27305-14_F09.EPS

Figure 9 Typical wheelbarrow, cart, and power buggy.

ing ability, making it suitable for repair work on many types of structures, for earthquake-proofing historical buildings, and for the construction of swimming pools and tunnels. It is also excellent for curves and special shapes.

There are dry-mix and wet-mix shotcrete processes. The original dry-mix process involves combining dry sand and cement in a mixer, then placing this mixture in a vertical double-chambered vessel. Under pneumatic pressure, the mixture flows through a rubber hose to the nozzle, where water, applied via a second hose, is combined with the material. The mixture leaves the nozzle under high velocity. The impact of the mortar on the receiving surface causes compaction.

Equipment used for the wet-mix process can involve an auger-type pump, in which plastic con-

crete is fed into the hose by means of a screw, then a high-velocity flow of air conveys the mixture to the nozzle where it is shot onto the receiving surface. Another type of equipment uses a pressurized tank in which rotating mixing paddles intermittently introduce air with the plastic concrete into the hose or pipe. The use of wet-mix shotcrete has an advantage over dry mix in that the amount of water can be established beforehand and maintained during the gunning operation.

2.3.0 Mixing and Patching Concrete Using Hand Tools

Hand and power tools can also be used for mixing and conveying concrete, as well as to repair and patch concrete. Although specialty craftper-

Cofferdam Concrete Seal Course

Cofferdams are temporary watertight structures used when building bridges, dams, and similar structures that keep water out of an excavation, such as the excavations for bridge piers. In order to remove the water from a cofferdam, the bottom of the cofferdam must be able to resist hydrostatic uplift. One method for doing this is to place a seal course of concrete at the bottom of the cofferdam structure before the water within the cofferdam structure is removed. The seal course is a concrete slab placed under water and constructed thick enough so that its weight is sufficient to counteract the force of the water trying to push its way through the bottom of the cofferdam.

The concrete for the seal course is placed through the water using a tremie pipe to supply concrete either through a hopper or by a concrete pump. Construction of the seal course is accomplished by first lowering the tremie pipe into position with a plug fitted into the pipe as a barrier between the water and concrete. Following this, concrete is charged into the tremie pipe. As the concrete is being placed, the end of the tremie pipe must always remain embedded in the concrete to a depth of at least 3 or 4 inches to prevent the water in the cofferdam from entering the concrete mix. Concrete placement and positioning of the tremie pipe must be done smoothly and deliberately. Also, the concrete mix must have good flow characteristics. Once concrete placement for the seal is started, it must continue without interruption until completion. For example, pouring of the seal course for one cofferdam during the construction of the Sidney Lanier Bridge in Georgia required 35 hours of continuous pouring.

Typically, a minimum of five days is allowed for the concrete to cure before attempting to remove the water from the cofferdam. For very deep cofferdams, removal of the water is sometimes done in stages. This allows for the installation of additional bracing at each stage to further support the walls of the cofferdam structure.

sons such as concrete finishers typically use these tools, there may be occasions when carpenters use these tools as well.

2.3.1 Combination Tools

Combination tools, such as those shown in *Figure 10*, are tools designed to perform specific jobs as indicated by the name of the tool. These tools can be made of metal or plastic.

2.3.2 Pointing and Margin Trowels

Pointing and margin trowels (*Figure 11*) are used for patching and mixing concrete. The pointing trowel is used to patch holes and to place grout at machine bases. The margin trowel is used mainly to mix concrete, patch holes, and to place grout at machine and steel column bases. It can also be used to finish patches in hard-to-reach areas.

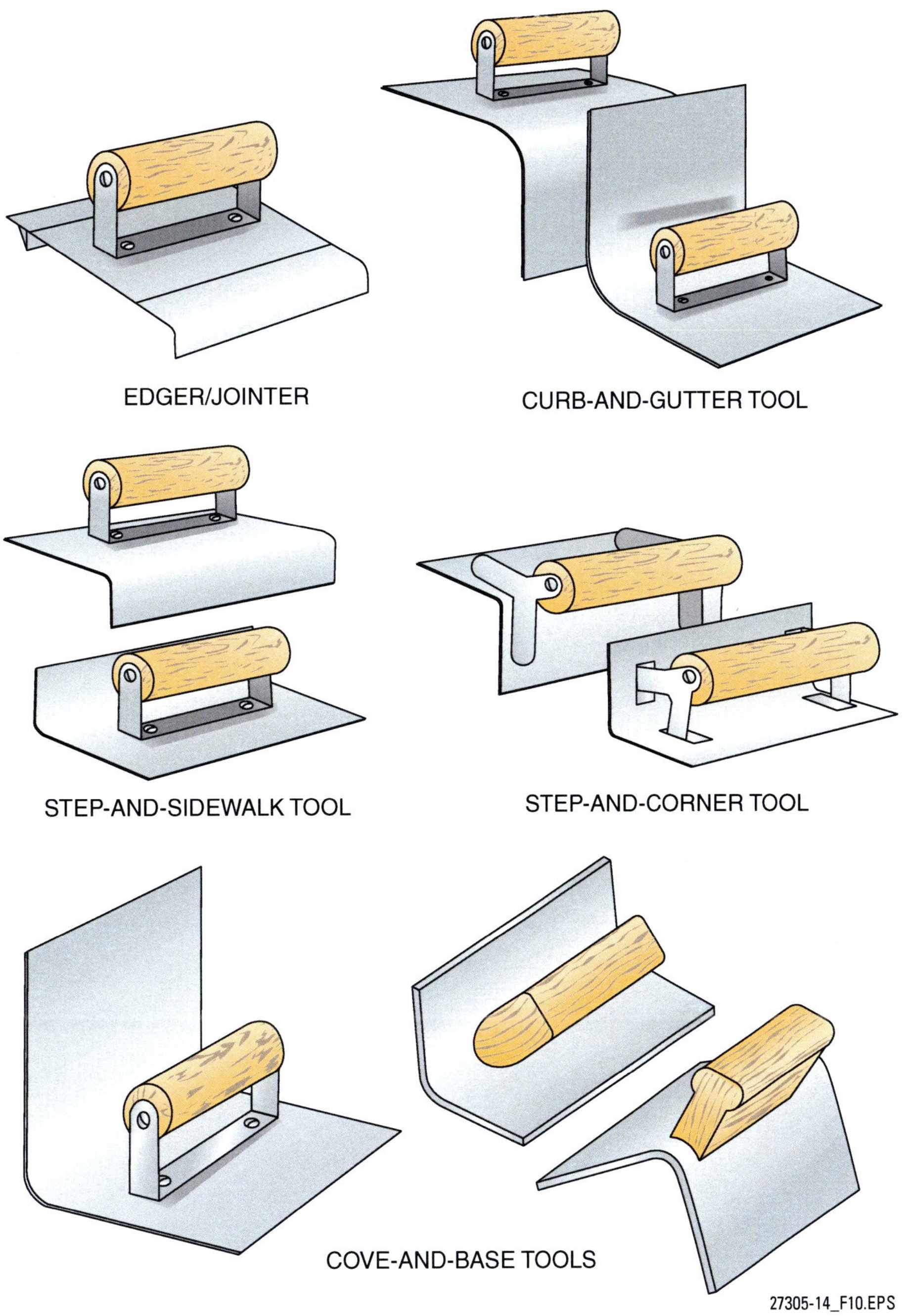

EDGER/JOINTER

CURB-AND-GUTTER TOOL

STEP-AND-SIDEWALK TOOL

STEP-AND-CORNER TOOL

COVE-AND-BASE TOOLS

27305-14_F10.EPS

Figure 10 Combination tools.

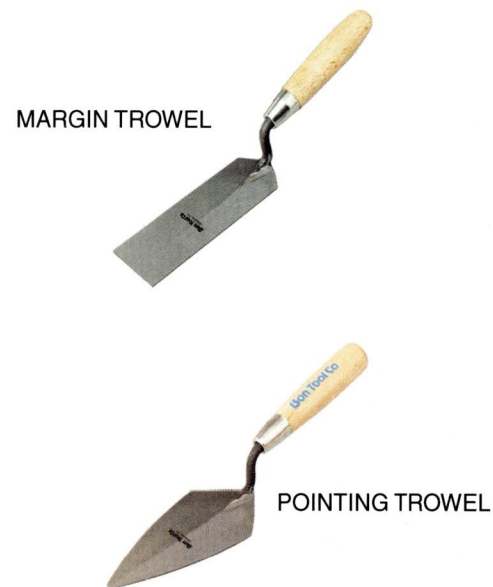

MARGIN TROWEL

POINTING TROWEL

27305-14_F11.EPS

Figure 11 Pointing and margin trowels.

Using Wheelbarrows

When using a single-wheel wheelbarrow, be careful not to overfill it with concrete, because tipping can easily occur with this type of wheelbarrow. To avoid this problem, it is better to use a two-wheel cart.

Gunite®

You may hear the term Gunite® used in connection with pressure-applied concrete. Gunite® is the original dry-mix-process concrete. It has been used extensively in the construction of in-ground swimming pools for many years. Dry-mix shotcrete and Gunite® are the same thing.

2.0.0 Section Review

1. The basic tasks involved in placing and finishing concrete include _____.

   a. segregating
   b. consolidating
   c. ordering
   d. mixing

2. Facilities that produce fully mixed plastic concrete are called _____.

   a. commercial mix plants
   b. batch plants
   c. central mix plants
   d. slump factories

3. Cement mortar sprayed onto a surface is called _____.

   a. shotcrete
   b. pneumatic concrete
   c. air-entrained concrete
   d. hosecrete

SECTION THREE

3.0.0 PLACING AND CONSOLIDATING CONCRETE

Objective

Explain the proper methods for placing and consolidating concrete into forms.
   a. Explain the proper method for placing concrete into forms.
   b. Explain the proper method for consolidating concrete.

Performance Task

Properly place and consolidate concrete in selected concrete forms.

Trade Terms

Bleed: A condition in concrete in which the solids settle and the water moves to the top.

Bulkhead: A vertical piece placed inside a concrete form to stop concrete at a certain point. A construction joint is created at the bulkhead.

Lift: Concrete layer of uniform thickness.

The task of placing concrete involves depositing the concrete in the forms, and then consolidating the concrete. Before placing the concrete, it is critical to ensure that the forms are ready.

3.1.0 Placing Concrete into Forms

Prior to concrete being placed into the forms, the formwork must be properly prepared to ensure it does not blow out. Formwork must be constructed and braced to resist the pressure of the concrete during placement and consolidation.

3.1.1 Preparing the Forms

Before concrete is placed in forms, the forms should be checked to make sure that they are accurately set, clean, tight, adequately braced, and have a form surface that will produce the desired appearance after the concrete is cured. All dirt, sawdust, shavings, tie wire, loose nails, and other debris should be removed from within the forms. The forms should be coated with a release agent such as form oil to prevent adhesion of the concrete and allow for easy removal. The release agent should be formulated for the particular usage and material to which it is to be applied. Always follow the safety precautions for form oil as specified on the manufacturer's safety data sheet (SDS).

All reinforcing steel should be cleaned. It should be free of mud, oil, loose rust, and mill scale. Loose, flaky, scaly rust that would affect the bond should be removed by using wire brushes or other stiff-bristle brushes. While following the proper safety precautions, grease and oil can be removed using a propane torch (being careful not to overheat the bars), or it can be washed off using a suitable solvent. If reinforcing steel is sticking out from a former placement of concrete, it should be cleaned of any dried mortar coatings or splashes that might have occurred during the former placement. Mortar that is so tightly bonded to the steel that it cannot be removed by vigorous wire brushing can remain. When the hazard is removed, impalement caps are no longer necessary.

Other points to check concerning form preparation include the following:

- The subgrade is firm and on grade, and the soil is undisturbed or heavily tamped
- Forms are properly aligned and braced
- Forms are set to proper grade
- Screeds are set at the proper location for efficient placing and leveling of the concrete
- Screeds are set to the proper grade
- **Bulkheads** are properly braced and set to the proper grade
- Reinforcing mesh or rods are properly placed and not touching the forms
- Expansion-joint material is in place and on grade
- Embedded items such as anchor bolts, traps, pipe, and conduit are held firmly in place using templates or by attaching them to the forms or reinforcing steel

On most commercial construction jobs, a formal final inspection of the formwork is required before placement of the concrete can occur. Such inspections typically involve representatives of the contractor, an inspector representing the owner, and various municipal building inspectors.

> **NOTE**
> If a formal inspection is required, never start placement of the concrete until the inspection has been performed and you have received written approval to proceed.

3.1.2 Placing the Concrete

Concrete-handling equipment used for placement should be clean and in good working condition. Any standby equipment should also be cleaned and ready for use in case of a breakdown.

When placing concrete (*Figure 12*), it should be deposited as near as possible to its final location in order to avoid excessive movement of the concrete. The concrete must be discharged from equipment at the proper flow rate. This allows the impact force to help in its placing and consolidation. Allowing concrete to dribble slowly causes it to pile up and the coarse aggregate to separate and roll down the sides.

To minimize separation in long drops, it is best to discharge concrete vertically, rather than at an angle. A short vertical drop chute should be used at the end of sloping chutes or conveyor belts. The related hopper should be filled by dropping concrete into the center, and it should discharge the concrete vertically from a center opening. The use of hoppers with side discharge or sloping gates should be avoided. If the concrete can be placed satisfactorily without the use of drop chutes or tremies, their use should be avoided. For example, if conditions allow, a crane bucket can be moved along the top of a wall form and concrete can be discharged directly into it.

27305-14_F12.EPS

Figure 12 Placing concrete from a chute.

When drop chutes must be used in order to prevent the concrete from striking the walls in narrow forms, they should be arranged so that they can be quickly moved and shortened, or a sufficient number should be supplied to cover the placement area without moving any drop chutes during the placement. The ends of tremies should remain vertical. Pushing them to the side results in a sloping condition and causes separation.

If required to place concrete on a sloping surface, start by placing the concrete at the bottom of the slope, then move up the slope. By using this method, the weight of the newly added concrete helps achieve proper consolidation. If using a chute for placement, a baffle should be placed at the end of the chute so separation is avoided and the concrete remains on the slope.

When placing concrete for a slab, the first discharge of the concrete should be placed along the edge of the form at one end, with each subsequent discharge placed on the heel of the fresh concrete already in place, not away from it. This method should be continued for the entire slab. Never dump concrete into separate piles or deposit it in one big pile, then move horizontally to fill a form. This practice results in segregation as the mortar flows ahead of the coarser aggregate. Concrete should be placed in a uniform thickness horizontally and must be thoroughly consolidated before the next deposit. The placement must be rapid enough so that the previous layer of concrete is still plastic when the new deposit is placed against it. This prevents flow lines and cold joints in the hardened concrete.

When placing concrete in wall forms, the concrete should be discharged from different positions along the form until a uniform layer, called a **lift**, has been placed in the form. This procedure is repeated until enough lifts have been placed to fill the form. For walls, beams, and girders, water should not be allowed to collect on the ends, in corners, or along the form faces. When concrete is placed in tall wall and column forms at a rapid rate, the concrete usually **bleeds**; that is, the solids settle and the water moves to the top. This is nor-

Estimating Quantity of Concrete

On some jobs, the carpenter may be responsible for determining how much concrete is needed to fill a particular form. For example, if the form for a basement floor slab is 25' × 32' × 4", how many cubic yards of concrete are needed to fill it?

---

mal. Do not use concrete of a stiffer consistency (lower slump) or slow down the placement to compensate for the gain in water.

3.2.0 Consolidating Concrete

After being placed in forms, concrete must be consolidated into a uniform solid mass within the form and around embedments and reinforcement. Proper consolidation prevents defects such as rock or stone pockets, entrapped air voids, and sand or gravel streaks from occurring in the concrete. Entrapped air is defined as an air pocket that is 1 millimeter or larger. Entrapped air should not be confused with the microscopic air bubbles incorporated in some concrete mixtures to improve their workability and resistance to cold weather conditions. Consolidation is not intended to remove the entrained air from such mixtures.

Consolidation can be performed manually or by using mechanical methods. For small concrete jobs with mixtures that are rather free-flowing, the concrete can be consolidated by hand using tamping rods, spades, shovels, or other suitable tools that will reach to the bottom of the lift. Thrusting up and down with these tools can accomplish the job without segregating the materials. Blows with a hammer applied to the outside of the form are also used to consolidate air pockets that can occur near the form walls.

The most frequently used method of consolidation is vibration (*Figure 13*). When concrete is vibrated, the friction between the coarse aggregate particles is temporarily destroyed and the concrete mixture flows like a liquid. The liquid concrete flows together in a compact mass due to gravitational forces and any entrapped air is released to the surface. As soon as the vibration ceases, friction is re-established. There are two basic types of vibration equipment: internal vibrators and external vibrators.

27305-14_F13.EPS

Figure 13 Using an internal vibrator to consolidate concrete.

3.2.1 Internal Vibrators

Internal vibrators, also called immersion, poker, or spud vibrators, are typically used to consolidate concrete in slabs, columns, beams, and walls. They consist of a flexible shaft with a vibrating head connected to a motor (*Figure 14*) that can be powered by electricity, gas, or air. Gas motors are often preferred and are specified on many jobs. There is an unbalanced weight inside the head that rotates at a high speed, causing the head to revolve in a circular motion.

The proper use of a vibrator is very important. Hit-or-miss penetration with a vibrator at all angles and without sufficient depth will result in improper consolidation between lifts of concrete. A systematic vibration of each new lift of concrete is necessary. The vibrator head should be lowered vertically into the concrete at regular intervals and allowed to descend by gravity to the bottom of the lift of concrete being placed and at least 6" into the preceding lift. For each insertion, the vibrator should be left in position for approximately five seconds and then slowly withdrawn. The length of time for immersion depends on the slump of the concrete. Watch for changes in the surface such as the appearance of a thin film of glistening paste and the escape of large bubbles of entrapped air to the surface.

Placing of Concrete in Cold and Hot Temperatures

Never place the concrete for a slab on a frozen subgrade because uneven settling and cracking of the slab usually results when the subgrade thaws. Prior to placing concrete for a slab in near-freezing temperatures, the subgrade should first be thawed and the reinforcing steel or welded-wire reinforcement warmed. The form, reinforcing steel, and embedded fixtures should be free of snow and ice when the concrete is placed.

The placing of concrete for a slab in temperatures above 85°F should also be avoided because the concrete sets much faster and may result in cracked or poorly finished slabs. When working in hot weather, avoid placing the concrete in the hottest part of the day. Start early in the morning such as at daybreak. It is also better to do the placement and finishing in smaller sections, if possible, and shade the area if practical.

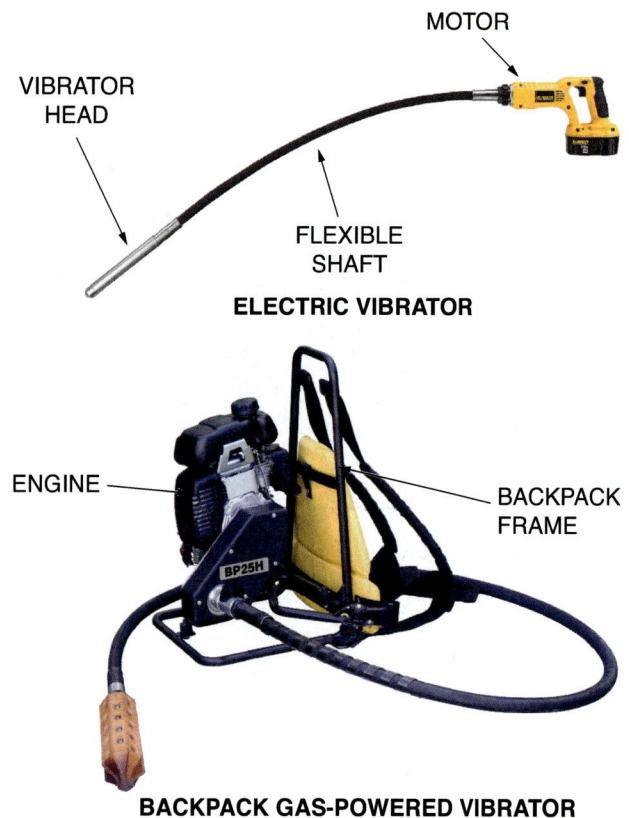

MOTOR

VIBRATOR
HEAD

FLEXIBLE
SHAFT

ELECTRIC VIBRATOR

ENGINE

BACKPACK
FRAME

BP25H

BACKPACK GAS-POWERED VIBRATOR

27305-14_F14.EPS

Figure 14 Electric- and gas-powered internal vibrators.

3.2.2 External Vibrators

There are a wide variety of external vibrators, including form vibrators, vibrating tables, and surface vibrators. Surface vibrators include vibrating screeds, plate vibrators, vibratory roller screeds, and vibratory hand floats or trowels. For most construction applications, internal vibrators are typically used. Vibrating screeds are sometimes used to both strike off and consolidate concrete when working with slab floors or other flatwork. More information about screeds, including vibrating screeds, is covered in the next section.

Use of Vibrators

To avoid segregation, internal vibrators should not be used to move concrete horizontally. The vibrator should not be allowed to come into contact with the form walls, as damage could occur to the architectural features and finishes built into the forms. Also, excessive vibration could blow out the form.

Consolidating Concrete in Column Forms

To help in the proper consolidation of concrete being placed into forms for support columns, access doors are sometimes built into the column forms at appropriate intervals along their height. These doors provide a convenient way to insert an internal vibrator into the form to consolidate the concrete as it is being placed progressively from the bottom to the top of the form. Always ensure the access door is properly closed and braced before placing concrete at that level.

3.0.0 Section Review

1. If safety rules are carefully observed, oil and grease can be removed from rebar by using _____.

   a. muriatic acid
   b. a propane torch
   c. sandblasting
   d. absorbent powder

2. The most frequently used method of consolidation is _____.

   a. incorporation
   b. segregation
   c. vibration
   d. fraternization

SECTION FOUR

4.0.0 FINISHING AND CURING CONCRETE

Objective

Describe the proper methods for finishing and curing concrete.

 a. Explain the proper method for screeding concrete.
 b. Explain the proper method for leveling concrete.
 c. Explain the proper method for finishing concrete.
 d. Describe how to properly cure concrete.
 e. Describe the use of joint sealants.
 f. Identify the tools used to rub and patch concrete.

Performance Tasks

Use a screed to strike off and level a concrete surface.

Use a bull float and/or darby to level and smooth a concrete surface.

Use an edger to form a radius at the edges of a concrete pad, slab, etc.

Use a hand float and finishing trowel to level high spots, remove imperfections, and smooth a concrete surface.

Trade Terms

Elastomeric: A material having the properties of excellent flexibility and elongation.

Spalling: The condition of concrete breakup (chipping, splitting, or crumbling).

While concrete finishing is generally left to concrete finishers or masons, carpenters can also be involved in this activity. This section focuses on the finishing of structural surfaces. The sequence for finishing structural concrete begins with screeding, then leveling, and finally finishing.

In warehouses and similar applications, good floor flatness and levelness is critical in order to allow higher and safer speeds for forklifts and other vehicles and to prevent shaking and vibration of such vehicles. Most commercial and industrial project specifications require that floors be measured for flatness and levelness in accordance with specification *ASTM E1155*. This specification assigns F_F (flatness) and F_L (levelness) numbers for these parameters, where the higher the F_F or F_L number, the greater the degree of flatness or levelness, respectively. The F_F number defines the maximum floor curvature over 24", computed on the basis of successive 12" elevation differentials. It is most affected by the finishing operation. Similarly, the F_L number defines the conformity of the floor surface to a horizontal plane as measured over a 10' distance. It is most affected by the forming and concrete screeding processes.

> **NOTE**
> These specifications are not applicable to elevated decks.

Specification *ASTM E1155* defines the requirements for minimum sample size, methods of calculating F_F and F_L numbers, methods for obtaining data points, and calculations for combining the results of multiple measurements to obtain overall F_F and F_L numbers. Floor flatness and levelness must be measured within 48 hours of concrete placement by a person with an American Concrete Institute (ACI) or other certification, using precise measuring instruments such as a rolling F-meter or dipstick floor profiler. Traffic patterns for floor-slab areas are categorized as being either a defined-traffic floor area or a random-traffic floor area. A defined-traffic floor area is one where forklifts and/or other wheeled vehicles travel in the same pattern at all times, such as in the aisles between racks in a warehouse. All areas not designated as defined-traffic areas are designated as random-traffic areas. Typically, the F-numbers specified for defined-traffic floor areas are higher than those specified for random-traffic floor areas. *Table 2* provides some examples of typical F-numbers for different floor-profile categories.

4.1.0 Screeding Concrete

Screeding is the first finishing process performed after the concrete is placed in the forms. It is the process of striking off the excess concrete to bring the top surface to the proper grade or elevation. Screeding removes the humps, fills the hollows, and results in a true, even surface. Depending on the size and type of job, screeding can be done manually or by using a powered screed.

A manual screed is a straightedge (*Figure 15*) that rests across the top of the form edges or across a pair of screed guides placed next to or within the form. Typically, two people, one on

---

Table 2 Example F-Numbers for Different Floor Flatness/Levelness Profiles

Floor Flatness/ Levelness Profile	Random-Traffic Floor Areas				Defined-Traffic Floor Areas
	Specified Overall Value		Minimum Local Value		
	F_f	F_L	F_f	F_L	F_{min}
Good	38	26	19	13	38
Flat	50	33	25	17	50
Very Flat	75	50	38	25	75
Superflat	100	66	50	33	100
Ultraflat	150	100	75	50	150

each side of the form, move the screed back and forth across the concrete in a sawing motion and advance it slightly with each movement. An excess of concrete should be carried along in front of the screed to fill low places as the screed is moved forward.

Manual screeds can be made of wood or magnesium. Common wood screeds are typically made of 1½" stock. Magnesium screeds are usually 1" thick and 4½" wide. Both types are available in lengths ranging from 3' to 16', with the length used depending on the distance between the edge forms or screed guides. Both wood and magnesium screeds can be used on the same types of jobs, but magnesium screeds work better on some concrete surfaces (toppings) because they prevent drag. Magnesium screeds are also preferred as a guide when cutting joints.

Powered screeds include small light-duty gasoline- or air-powered screeds, heavy-duty screeds, and laser screeds. Small gasoline- and air-powered screeds (*Figure 16*) function to both consolidate and smooth the concrete. These screeds typically have interchangeable magnesium blades of varying lengths for use on different size slabs. Most are operated by either one or two workers. Heavy-duty vibrating and roller screeds (*Figure 17*) are used to strike and consolidate wide expanses of high-vol-

ume concrete. They can be self-propelled or pulled by crank winches. Both types have an attached method of vibration and a metal or metal-covered straightedge, which can be adjusted to compensate for sag or designed crown or slope. Because of their heavy weight, steel truss screeds must be supported by side rails or side forms. They also require a relatively long setup time.

When working with slabs that require very flat profiles, self-propelled laser screeds, similar

27305-14_F16.EPS

Figure 16 Small one-man-operated gasoline-driven power screed.

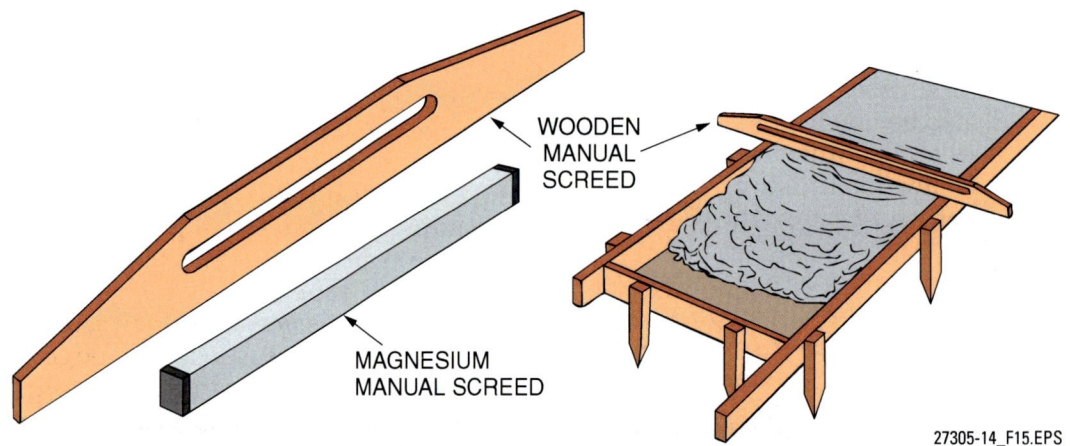

WOODEN MANUAL SCREED

MAGNESIUM MANUAL SCREED

27305-14_F15.EPS

Figure 15 Typical manual screeds (straightedge).

NCCER – *Carpentry Level Three* 27305-14

27305-14_F17.EPS

Figure 17 Roller screed.

to the one shown in *Figure 18*, are widely used. Laser screeds have a hydraulic-powered laser-controlled screed/compacting head mounted on a telescopic boom. The self-leveling screed head consists of a plow that removes excess concrete, an auger that cuts concrete on grade, and a vibrator that consolidates the concrete material. Depending on the manufacturer and model, they have different screed head widths and boom length extensions to satisfy different job size requirements. Laser receivers mounted at each end of the screed head receive a signal several times per second from a related transmitter located at the proper floor elevation level to provide totally automatic control of the finished floor level.

Use of a laser screed involves placing concrete in strips to match the size of the screed head at a depth of about 1" higher than final grade. Following this, the laser screed is moved into position and the boom extended over the placed concrete. The screed/compacting head is then lowered to the established grade as controlled by the laser transmitters. Retraction of the boom causes the screed head to be drawn across the fresh concrete, which is leveled and compacted in a single pass. Once a pass is completed, the machine is repositioned to the right or left of the previously screeded area, and with some overlap, the operation is repeated.

27305-14_F18.EPS

Figure 18 Laser screed.

Before a vibratory screed is used on flatwork, an internal vibrator should be used along the edges of the form. This is because surface vibration is least effective in these areas. Also, vibrating screeds should not be used on concrete with slumps in excess of 3". Surface vibration of such concrete will result in an excess accumulation of mortar and fine material on the surface and thus reduce wear resistance. For the same reason, surface vibrators should not be operated after the concrete has adequately consolidated.

4.2.0 Leveling Concrete

Immediately after screeding and before the free water can bleed to the surface, the concrete should be further leveled using a darby or long-handled bull float. This process eliminates high and low spots in the concrete and embeds the large aggregate just below the surface. It also brings sufficient mortar to the surface in preparation for other finishing.

Darbies (*Figure 19*) are long, flat, rectangular pieces of wood or magnesium that have handles. They are available in lengths from 2½' to 6½' and widths from 3" to 4". Both can be used on the same types of jobs, but magnesium darbies are more durable, prevent drag, and are preferred on some toppings and on air-entrained concrete. When using a darby, sweep the tool in wide arcs while applying a light pressure on the blade's trailing edge. Work from the center to the edges. When using a darby on wide slabs, it is necessary to support yourself on knee boards so that you can reach all areas without damaging the concrete.

Bull floats (*Figure 20*) are more commonly used to float concrete in locations where there is enough room to accommodate their 4'- to 16'-long handles. The tool itself is a large, flat, rectangular piece of wood or metal (aluminum or magnesium) that is usually 8" wide and ranges in length from 3½' to 5'. Both wood and magnesium bull floats can be used on most floor slabs, pavement, and sidewalks. However, magnesium bull floats prevent drag, are much lighter, and are preferred on certain toppings, low-

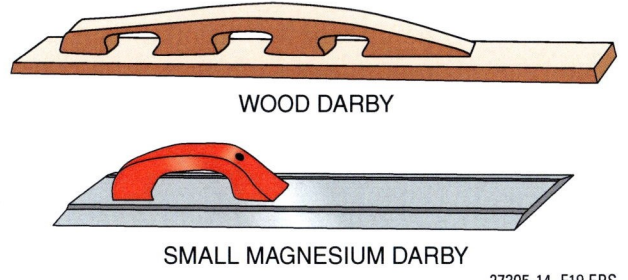

WOOD DARBY

SMALL MAGNESIUM DARBY

27305-14_F19.EPS

Figure 19 Darbies.

27305-14_F20.EPS

Figure 20 Bull floats.

27305-14_F21.EPS

Figure 21 Rollerbug® tamper.

slump concrete, and air-entrained concrete. When using the bull float, the tool is pushed across the concrete at right angles to the screed marks with the front edge slightly raised, then it is pulled back with the blade flat.

When working with low-slump concrete mixes (1" or less), it is sometimes necessary to use a tamper to consolidate the concrete and settle the larger aggregate just below the surface of the concrete. One type of tamper used for this is the 3' to 6' Rollerbug® (*Figure 21*). The size of the job determines the size of the tamper used. When using a Rollerbug® tamper, do not overwork the concrete as this will force the coarser aggregate too far down below the surface, resulting in a weaker surface condition. For some jobs, no further finishing will be required. Other jobs may require one or more additional finishing operations such as edging, jointing, troweling, and brooming.

> **NOTE**
> Before using a Rollerbug® tamper on a project, make sure that its use is permitted. Some job specifications prohibit the use of Rollerbug® tampers.

4.3.0 Finishing Concrete

Generally, finishing of the concrete surface begins after the concrete has been allowed to harden enough to support the weight of an average person, leaving only slight footprints on the surface. The bleed water and water sheen or glossy appearance should be gone. Finishing can involve the following tasks:

- Edging and jointing
- Floating
- Troweling
- Brooming

4.3.1 Floating

After the concrete has been edged and jointed, it should be floated. Floating further embeds the aggregate particles just beneath the surface, removes slight imperfections, compacts the mortar at the surface, and keeps the surface open so excess moisture can escape. Floating produces a relatively even (but not smooth) texture. It is often used for a final nonslip finish, especially for exterior slabs. When a nonslip finish is desired, it may be necessary to float the surface a second time. Marks left by edgers and hand jointers are normally removed during floating unless they are desired for decorative purposes. If a decorative edge is desired, those tools should be rerun after final floating. As with all finishing operations, the surface should not be overworked while

Leveling

The concrete should not be overworked during the leveling process. This can cause excessive bleeding to occur, thus weakening the concrete surface.

it is still plastic as this will bring an excess of water and fine materials to the surface and result in surface defects.

Floating can be done using hand floats moved in a circular fashion to smooth out bull-float or darby marks. On larger areas, support yourself on knee boards. If desired, you can also lean on a second float for additional support. Common hand floats (*Figure 22*) are made of wood or metal (steel, aluminum, or magnesium). Metal floats are usually 3½" wide and made in two lengths, 12" or 16". Wood floats are made in 12", 15", or 18" lengths and in 3½" or 4½" widths. Metal floats prevent drag, are lighter than wood, and are preferred on some types of toppings and on air-entrained concrete. Special types of hand floats made of cork, rubber, sponge, or carpet are also available to create textured finishes on rubbed concrete.

Floating can also be done mechanically using powered finishing machines (*Figure 23*). These machines can be used to both float and trowel a surface by equipping it with the applicable float or trowel blades. Single-rotary machines are commonly used for smaller jobs. These typically have three blades and are controlled by an operator walking along with the machine (*Figure 24*). Operator-driven multiple-rotary machines with three rotaries are used mainly for finishing jobs involving very large areas of concrete. Note that hand floating is still required around openings and other obstructions that cannot be finished mechanically.

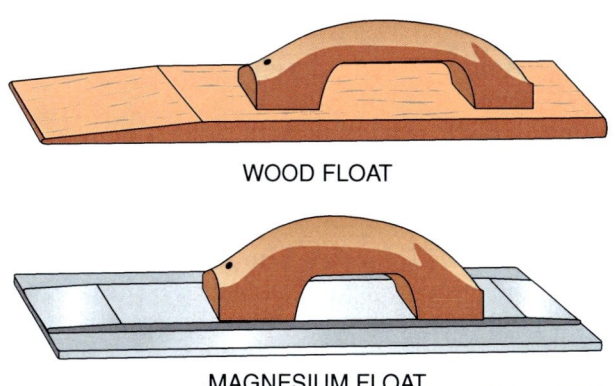

WOOD FLOAT

MAGNESIUM FLOAT

27305-14_F22.EPS

Figure 22 Hand floats.

27305-14_F23.EPS

Figure 23 Powered finishing machine.

> CAUTION
>
> Craftworkers must be properly trained on the safe use of powered finishing machines to ensure their own safety and the safety of others working around the equipment. When working around powered finishing machines, always be aware of their location and stay clear of the area until the equipment is stopped or moved to another area.

4.3.2 Troweling

Where a smooth, hardened surface is desired, the floating process is followed by troweling. Troweling should be delayed as long as possible, but not so long that the surface becomes too hard to finish. Typically, troweling can start when the surface is hard enough to leave only a fingerprint. Starting prematurely may cause scaling and dusting of the concrete. As with floating, troweling can be done manually or with a finishing machine equipped with troweling blades.

Hand trowels (*Figures 25* and *26*) are made from either high-carbon tempered steel or specialized

27305-14_F24.EPS

Figure 24 Using a single-rotary finishing machine.

long-wearing stainless steel. They are available in a variety of sizes ranging from 10" to 20" long and from 3" to 5" wide. The size used depends on the job. For example, a long trowel is normally used for the first troweling process, while a shorter trowel is used for final troweling.

During the first troweling, the trowel should be held flat against the surface. If it is tilted or pitched at too great an angle, an objectionable washboard or chatter surface will result. The first flat troweling may produce an adequate surface that is free of defects, but additional troweling with the blade tilted may be necessary to increase the smoothness and hardness. There should be a time delay between trowelings to allow the concrete to become harder. As the surface hardens, each successive troweling should be done with a smaller-sized trowel to allow enough pressure to be exerted to increase the density, strength, and hardness of the surface. The final pass should make a ringing sound as the tilted blade moves over the hardening surface. Again, when hand-troweling large slab surfaces, you must support yourself on knee boards.

When the first troweling is done by machine, at least one additional troweling by hand is required to remove small irregularities. If necessary, tooled edges and joints should be rerun after troweling to maintain uniformity and true lines.

MIDGET TROWEL

PIPE TROWEL

27305-14_F25.EPS

Figure 25 Hand trowels.

27305-14_F26.EPS

Figure 26 Using hand trowels.

4.3.3 Edging and Jointing

Edging is required to round the upper corners of concrete at the form to prevent spalling of the edges and to improve the appearance. Edging compacts the concrete next to the form, where floating and troweling are least effective.

The concrete should be cut away at the form about an inch deep using an aluminum float or a pointed trowel, as shown in *Figure 27*, and then finished with an edger (*Figures 28* and *29*). Edgers are made of steel, stainless steel, bronze, or malleable iron. They are available in many sizes, but the most common sizes are those from 6" to 10" long, from 1½" to 4" wide, with a lip that is from ⅛" to ⅝", and with a radius from ⅛" to 1½". Some edgers have a long handle so that they can be used from a standing position. When using an edger, it must be held flat with the front raised slightly to prevent digging into the surface.

Either during or immediately following the edging operation, control joints should be made in the slab, sidewalk, driveway, or other horizontal flatwork, to induce cracking at desired points rather than at random points. This can also be done in the green state using a power saw. Control joints compensate for contraction of the concrete caused by drying shrinkage.

27305-14_F27.EPS

Figure 27 Cutting away a concrete edge using an aluminum float.

Making Control Joints

In order to make control joints straight and square to the concrete slab, use a straight 1 × 8 or 1 × 10 board of appropriate length as a guide. Prior to making the control joint, make sure that the board is placed square with the slab on both sides of the form.

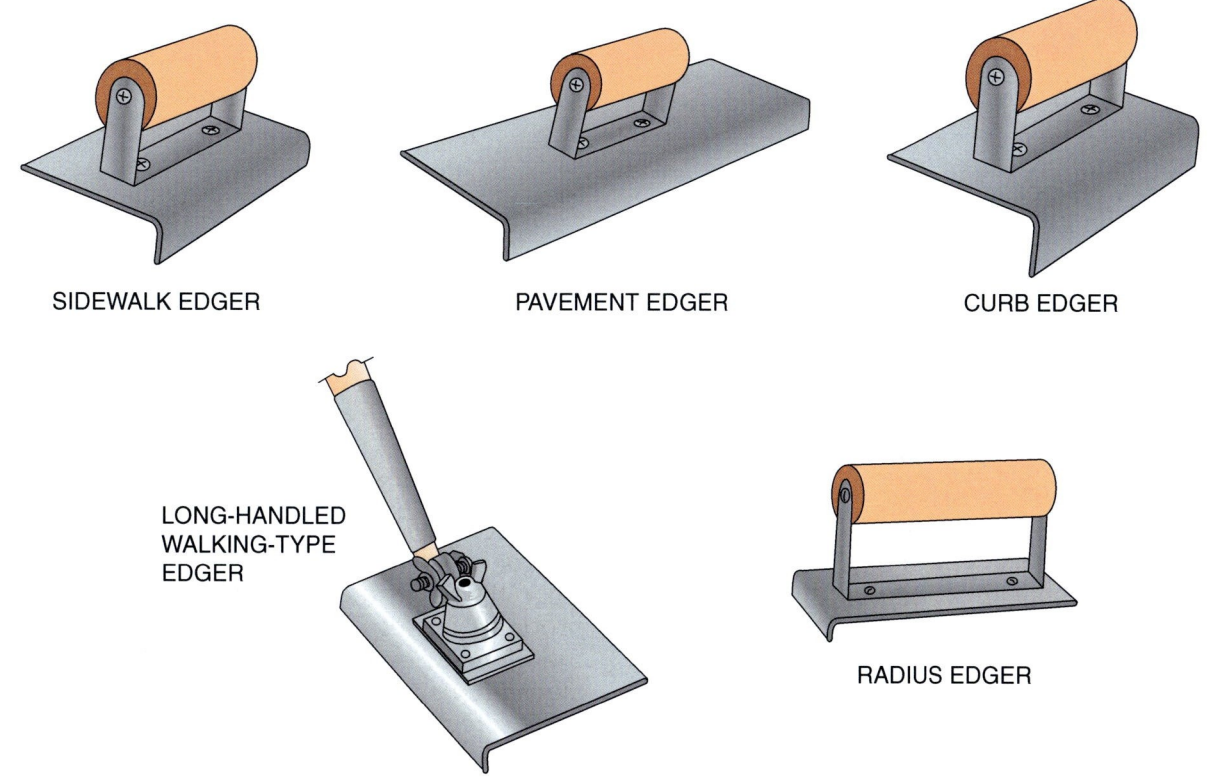

SIDEWALK EDGER

PAVEMENT EDGER

CURB EDGER

LONG-HANDLED WALKING-TYPE EDGER

RADIUS EDGER

27305-14_F28.EPS

Figure 28 Edgers.

27305-14_F29.EPS

Figure 29 Using an edger.

The depth and distance between control joints depend on the size, location, and depth of the slab. Typically, the depth of the control joint should be at least one-fifth to one-quarter the thickness of the slab and a minimum of 1" for all slabs. The rule of thumb for spacing control joints in a slab is to space joints, in feet, two to three times the slab thickness, in inches. To apply this rule to a 4" slab: Doubling 4 = 8, and tripling 4 = 12. Therefore, the control joints should be spaced 8' to 12' apart.

Control joints can be made using a hand jointer (groover) or power saw, or by inserting strips of plastic, hardboard, wood, metal, or preformed joint material into the plastic concrete.

Hand jointers (*Figures 30* and *31*) are made of stainless steel, bronze, or malleable iron. Common jointers are 6" long and vary in width from 2" to 4½". They have a shallow, medium, or deep bite (cutting edge) ranging from $\frac{3}{16}$" to 1" in depth. Some jointers also have a long handle so that they can be used while the finisher is standing.

4.3.4 Brooming

A slip-resistant surface can be produced by brooming the surface of the concrete (*Figure 32*). This should be done before the concrete has thoroughly hardened, but while it is sufficiently hard to retain the scoring. Rough scoring can be achieved with a rake, steel-wire broom, or stiff, coarse-fiber broom. Brooming is usually done af-

Control Joints in Sidewalks and Driveways

How far apart would you space control joints when constructing a concrete sidewalk or driveway?

**LONG-HANDLED
WALKING-TYPE JOINTER**

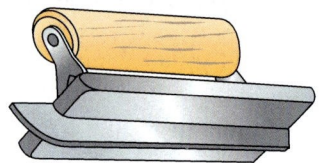

**SMALL-BITE
HAND-TYPE JOINTER**

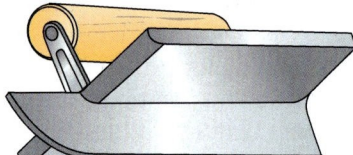

**DEEP-BITE
HAND-TYPE JOINTER**

27305-14_F30.EPS

Figure 30 Jointers (groovers).

27305-14_F31.EPS

Figure 31 Using a hand jointer.

27305-14_F32.EPS

Figure 32 Brooming a concrete surface.

ter floating. However, if a finer texture is desired, the concrete should be floated and troweled to a smooth surface, then brushed with a soft-bristled broom. Best results are obtained with a broom that is made specifically for texturing concrete. Slabs are usually broomed perpendicular to the main direction of traffic.

4.4.0 Curing Concrete

Once concrete has been placed and finished, care must be taken to make sure it cures properly. Hydration is a chemical reaction that takes place when water is combined with cement, sand, and gravel in a concrete mixture. It is this chemical reaction that causes concrete to harden. Curing concrete is the process of retaining the moisture in freshly placed concrete long enough to allow for proper hydration to take place.

Concrete must be protected against moisture loss during the early stages of hydration in order for it to cure properly. If the water contained in the concrete is allowed to evaporate too quickly, hydration of the cement will stop, and there will be no further gain in the strength and durability of the concrete. As concrete dries, it shrinks, and if drying occurs when the concrete has little strength, cracks will result. In addition, since drying occurs first on the surface, the cement will not be hydrated there but will be present as a dust coating with no strength to hold the aggregate together.

The degree and speed of concrete hardening depends on many factors such as the amount of cement in the mixture, type of cement, temperature of the concrete, temperature of the surrounding air, and chemical admixtures used in the concrete mix. For standard concrete mixtures, the first three days after placing are the most critical. During this time, concrete is most exposed to damage. At seven days, it reaches about 70 percent of its strength and at two weeks, it is at about 85 percent. Under normal conditions, maximum strength is reached at about 28 days.

The greatest influence on the curing rate of concrete is the temperature. Higher temperatures speed the curing rate. Proper hydration occurs when the temperature ranges between 55°F and 73°F. Above 73°F, the hydration process should be slowed to gain the greatest strength of the concrete. Some methods that are commonly used to slow the hydration process include the following:

- Adding water-reducing admixtures to the concrete mixture
- Spraying the forms, reinforcing steel, and subgrade with cool water immediately before placement of the concrete
- Spraying a curing agent or compound on the concrete surface to form an impassable film that prevents or retards the escape of moisture from the concrete (*Figure 33*)
- Covering the concrete surface with a curing blanket of wet burlap (*Figure 34*), polyethylene sheeting, or waterproof paper to prevent rapid evaporation of the moisture
- Spraying the finished surface continuously with a water mist using a system of water pipes and spray heads

Hydration takes place at a slower rate when temperatures are low. When concrete temperatures fall below freezing, no hydration takes place and the concrete can be permanently damaged. To compensate for this condition, the concrete mixture is usually heated both during and after placement. Using a concrete mixture with air-entraining or accelerating admixtures is another method used to aid curing when placing concrete in cold weather.

Wind, rain, and traffic on a slab are some other factors that must be considered when curing concrete. Winds affect curing by speeding the evaporation of water. The combination of hot, dry weather and wind can dry water from the concrete surface faster than bleed water can replace it. Strong wind can speed evaporation in cold weather as well. In strong wind, surfaces dry faster, so scheduled wetting needs to occur more frequently. Wind can shift water from sprayers, sprinklers, and foggers away from the concrete. It can unseal lap joints and edge seals of plastic sheeting or impervious paper. Strong wind can lift and tear unsecured plastic sheeting or paper. Windbreaks can be built out of tarpaulins to protect fresh concrete from evaporation.

Rain creates problems by adding too much moisture. Rain on freshly placed concrete can erode the surface and dilute the cement paste at and near the surface. Rain can wash away fines and cause voids in concrete being placed. Finishing of unformed concrete cannot be completed in the rain because it will be damaged. Freshly placed concrete must be sheltered completely from rain, or work stopped until the rain is over. Temporary rain shelters can be built from tarpaulins or plastic sheeting to protect concrete surfaces. Check that the rain will not drip off the edges of the shelter onto the slab.

As concrete cures, it gains strength over time. During this time, construction traffic and loads driven over new slabs can damage the surface, can exceed the strength of the slab, and can lead to cracked or damaged slabs. The best protection is to keep traffic off the slab for the first seven days. If activity must continue, it should start with smaller equipment, foot traffic, and light material loads. This traffic is acceptable after three to seven days, the initial cure time. The loads must be light, and

27305-14_F33.EPS

Figure 33 Spraying a concrete surface with curing compound.

27305-14_F34.EPS

Figure 34 Wet burlap blanket on a concrete surface to retain moisture.

the concrete surface must be protected. Kraft paper, plywood, or fiber sheets can be used to provide traffic paths over a slab. Care should be taken to prevent damage to edges by routing paths away from edges and by reinforcing path material at the edges of the slab. Concrete without reinforcement is strong in compression (crushing force) but not in tension (pulling force) or flexure (bending force). Loading or heavy traffic can bend the concrete, especially at the edges of slabs, and cause cracks. Use common sense to protect new concrete from loads. Residential slabs, driveways, and walks are not designed to carry heavy loads such as ready-mix trucks or heavy materials. These types of loads should not be allowed on residential concrete.

4.4.1 Sprayers

Sprayers (*Figure 35*) are used to apply form oil to concrete forms, to spray curing compounds, and to mist or fog water over finished concrete to reduce rapid evaporation.

4.5.0 Joint Sealants

Sometimes the concrete finishing process requires that joint sealant be applied to the joints in the finished concrete structure. Joint sealants are flexible materials used to seal construction, isolation, and similar joints subject to movement between adjacent sections of concrete or between concrete and other construction materials.

A wide variety of sealant formulations are available for use on most interior and exterior surfaces that require sealing, including concrete, masonry, metal, glass, ceramic, and wood. Their durability, flexibility, and other features vary, depending on how they are formulated. Different applications require different types of sealant. Polyurethane and polysulfide sealants are widely used in commercial/industrial applications for sealing expansion and control joints in precast concrete panels and on joints in sidewalks, parking lots, terrace decks, etc. They are formulated for exterior use as highly elastomeric materials. This means they have excellent flexibility and elongation properties. They come in one- and two-part formulations, and have excellent resistance to weather, ultraviolet light, and moisture. They cure quickly with little shrinkage and stay flexible for years in most environments.

The requirement for the use of joint sealants and the type of sealant to use are normally specified in the construction drawings and/or specifications for the structure. If sealants are required but not specified, follow the sealant manufacturer's recommendations when selecting a sealant for a particular job.

27305-14_F35.EPS

Figure 35 Typical concrete sprayer.

Regardless of the type of sealant used, the joint must be prepared properly in order for the sealant to bond correctly. Joint preparation must be done in accordance with the manufacturer's instructions for the sealant. Generally, this involves making sure that the joint is dry and free of oil, dust, or other debris. Joints cast in new concrete can contain form oil, loose mortar, and dust. They can be cleaned by sandblasting, power and/or hand wire brushing, or by using power-driven routers. Sawed joints will be dusty and should be blown out with oil-free air. If honeycombs or faults exist on the side of the joint slot, they should be patched before application of the sealant.

4.6.0 Rubbing and Patching Concrete

Concrete may need to be trimmed to reduce high spots or smoothed after it has hardened. Hand or power tools may be used to perform this activity.

4.6.1 Hammers

There are two types of hammers used by finishers to work with hardened concrete: the chipping hammer and the bush hammer (*Figure 36*).

CHIPPING HAMMER

BUSH HAMMER

27305-14_F36.EPS

Figure 36 Cement hammers.

Chipping hammers are used to cut off projections and high places on hardened concrete. The bush hammer is used to fracture aggregate to present a rough, decorative finish.

4.6.2 *Carborundum Rubbing Stones*

Carborundum rubbing stones (*Figure 37*) are used to smooth partly hardened concrete. Common rubbing stones are 6" to 8" long, 2" to 3½" wide, and ¾" to 2" thick. Grit sizes range from No. 24 (coarse) to No. 150 (fine).

4.6.3 *Power Grinders*

Power grinders (*Figure 38*) are used to remove seams and projections, reduce high spots, and smooth surfaces. Wear appropriate personal protective equipment when using power grinders. Electric handheld grinders are used for smaller areas. Larger gasoline-powered units are used for larger areas. Single- and double-head concrete grinders are available in various sizes. Scarifiers are used for rough finishing.

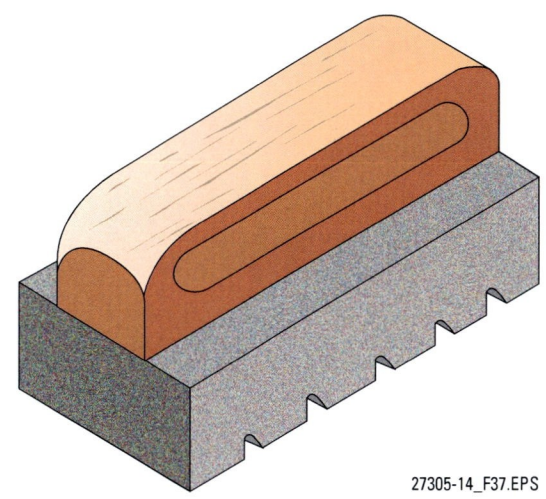

27305-14_F37.EPS

Figure 37 Typical carborundum rubbing stone.

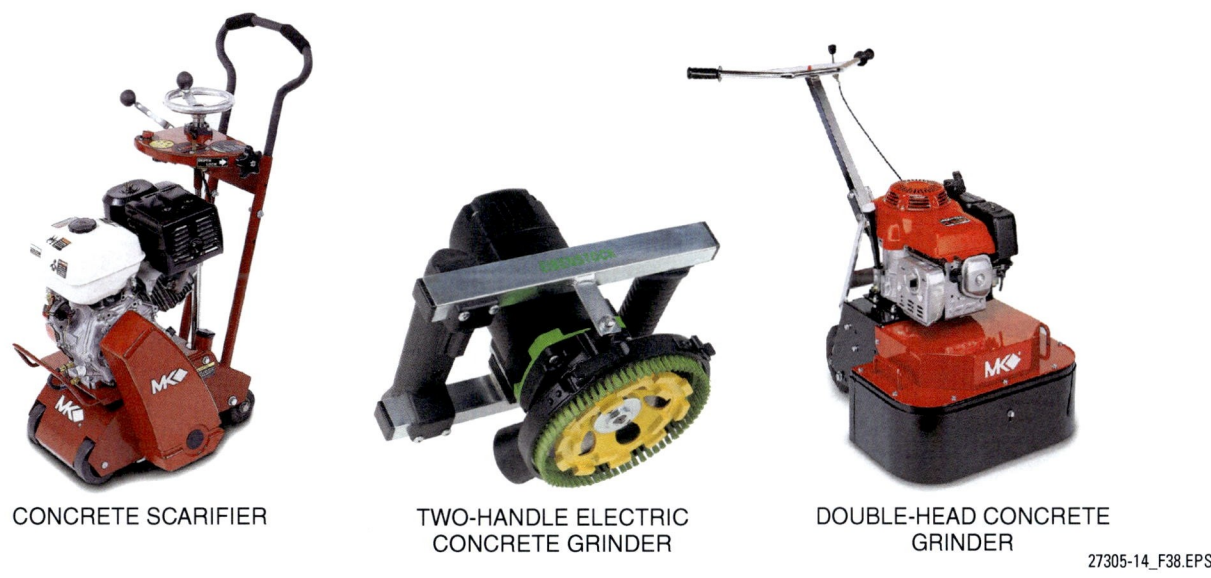

CONCRETE SCARIFIER

TWO-HANDLE ELECTRIC
CONCRETE GRINDER

DOUBLE-HEAD CONCRETE
GRINDER

27305-14_F38.EPS

Figure 38 Power grinders.

4.0.0 Section Review

1. To meet ASTM specifications, floor flatness and levelness measurements after concrete is placed must be done within _____.

   a. 24 hours
   b. 36 hours
   c. 48 hours
   d. 72 hours

2. Bringing the top surface of the concrete to the proper elevation is done with a tool called a _____.

   a. strike
   b. screen
   c. leveler
   d. screed

3. A bull float may have a handle as long as _____.

   a. 6 feet
   b. 12 feet
   c. 16 feet
   d. 20 feet

4. Rounding the edges of concrete at the form improves appearance and prevents _____.

   a. segregation
   b. spalling
   c. delamination
   d. cracking

5. The greatest influence on the curing rate of concrete is the _____.

   a. temperature
   b. concrete volume
   c. water/cement ratio
   d. humidity

6. The term used to describe a material that has excellent flexibility and elongation properties is _____.

   a. malleable
   b. ductile
   c. elastomeric
   d. tensile

7. Hydration rate is *not* affected by the surrounding air temperature.

   a. True
   b. False

5.0.0 CONCRETE JOINTS

Objective

Identify the different kinds of joints in concrete structures.

    a. Identify construction joints.
    b. Identify isolation joints.
    c. Identify control joints.
    d. Identify decorative joints.

Performance Task

Use a jointer to make control joints in a concrete surface.

Trade Terms

Laitance: Fine particles on a concrete surface resulting from an upward movement of water in the concrete.

Set: The hardening of concrete.

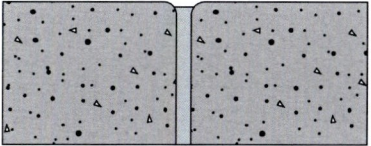

TOOLED CONSTRUCTION JOINT
(WITH EXPANSION MATERIAL)

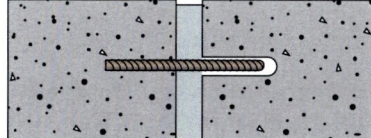

CONSTRUCTION JOINT
(WITH STEEL DOWEL AND SLIP CAP)

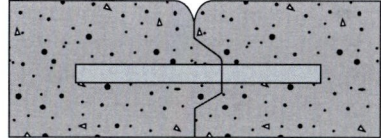

TOOLED CONSTRUCTION JOINT WITH
KEYWAY AND SMOOTH STEEL DOWEL

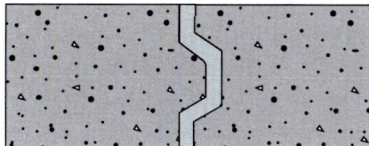

CONSTRUCTION JOINT
(WITH KEYED EXPANSION)

27305-14_F39.EPS

Figure 39 Types of construction joints.

It is important to become familiar with the different kinds of concrete joints and the specific purpose for each type of joint. Joints are made in concrete structures to do the following:

- Control cracking of concrete due to temperature and moisture changes
- Allow movement of foundations or building components
- Increase tensile strength
- Add to the decorative element of concrete

Joints can be grouped into four classes: construction joints, isolation joints, control joints, and decorative joints.

5.1.0 Construction Joints

Construction joints are formed at the point where concrete placement is temporarily stopped. An example of a construction joint is the joint produced by a bulkhead to break up a large floor area into reasonably sized areas. Another example is the joint formed between succeeding layers of concrete placed in a wall form. *Figure 39* shows some common methods used for forming construction joints.

Ideally, the use of construction joints should be kept to a minimum because they can be a source of water leaks. When dealing with construction joints, take precautions to ensure that a good bond will exist between one layer of concrete and the next.

To achieve a good bond between fresh concrete and previously placed concrete at a construction joint, the previously placed concrete must be properly prepared. This is especially important if the joint is to be watertight and durable. The surface of the previously placed concrete should be cleaned so that it is free of **laitance**, dirt, oil and grease, curing compound, soft concrete, and other debris. The surface should be slightly roughened. In addition to wire brushing, wet sandblasting and high-pressure power washing are two methods commonly used for cleaning a construction joint. A properly prepared joint will have a clean, clear, sharp appearance similar to that of a fresh break in sound concrete. To achieve a good bond in a construction joint, proceed as follows:

- Clean the surface of the first layer of concrete carefully. Wire-brush the concrete to expose the coarse aggregate before the first layer has become thoroughly hard.

- Dampen the surface of the first layer of concrete.
- Apply a coat of portland cement grout on the surface of the first concrete layer. Bonding agents are also available for this purpose. You can then begin placing the next layer of concrete.

5.2.0 Isolation Joints

Isolation joints (*Figure 40*), also called expansion joints, are used to separate different parts of a structure to permit both vertical and horizontal movement. For example, isolation joints are used to separate a new concrete slab from adjoining building materials or from older cured concrete. They are also used around the perimeter of a floating floor slab and around column piers and other isolated foundation forms. Isolation joints allow new concrete to expand and contract at its own rate, unaffected by the different curing rates of the adjoining materials.

Isolation joints are created prior to the concrete being placed by installing ¼" to ½" premolded filler strips, typically made from asphalt-saturated fiber or similar material, at the required locations. The strips should extend for the full thickness of the slab and can be placed so that the top edge is flush with the finished level of the slab, where no safety hazard from tripping exists. If the strips are not installed flush with the finished level, the joint is typically sealed with joint-sealing compound. The premolded strips remain permanently in place after the concrete has set. The strips can be placed at grade level to establish the screed line.

5.3.0 Control Joints

Control joints, also called contraction joints, are used in a concrete structure to induce cracking at desired points rather than at random points. Control joints compensate for contraction of the concrete caused by drying shrinkage. If control joints are not used in slabs or lightly reinforced walls, random cracks will occur. *Figure 41* shows some examples of different kinds of control joints. Control joints can be made using a hand groover or they can be sawed using a power saw equipped with a concrete blade. Other methods used to make control joints include placing plastic strips, steel T-bars, or polyethylene strips in the concrete.

For slabs, the depth of control joints should be at least one-fifth to one-quarter the thickness of the slab and a minimum of 1" for all slabs. Manual tooling of control joints requires that the hand grooving tool have sufficient depth required by the thickness of the slab. If a saw-cut joint is made in a slab, it must be sawed into the concrete after it is sufficiently hardened, but before the slab starts to crack. To avoid problems with cracking, some concrete finishers make the joints with a hand groover first, then cut them deeper with a saw later.

5.3.1 Cutting Control Joints

If a saw-cut joint is made in a slab, it must be sawed into the concrete with a power saw (*Figure 42*) after

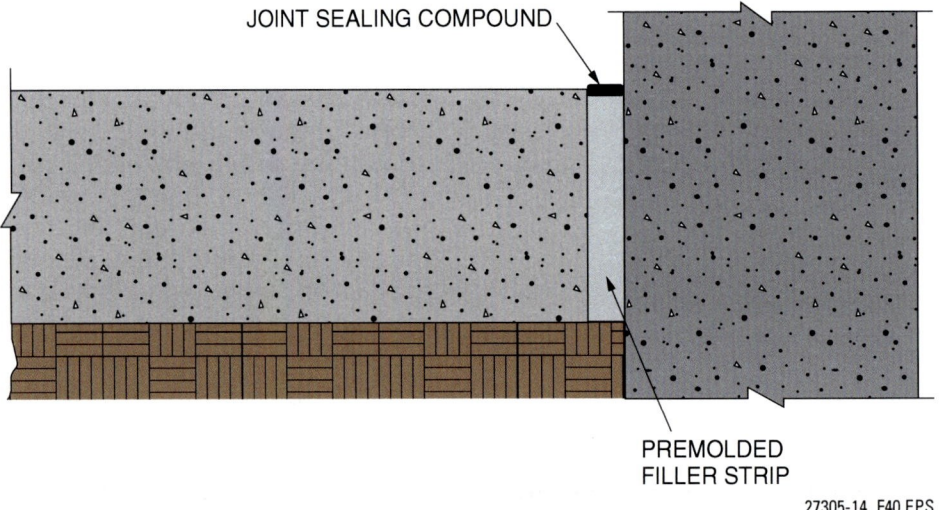

Figure 40 Isolation joint.

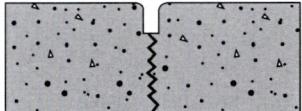

TOOLED CONTROL JOINT

SAWED CONTROL JOINT

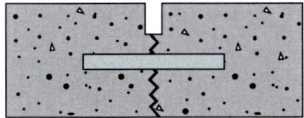

SAWED CONTROL JOINT
(WITH SMOOTH STEEL DOWEL)

27305-14_F41.EPS

Figure 41 Types of control joints.

it is sufficiently hardened to prevent raveling of the saw cut, but before the slab starts to crack. *Figure 43* shows a typical floor-type power saw that can be used for both dry and wet sawing of concrete. It has a built-in water distribution system that supplies water to both sides of the saw blade for wet sawing. When wet sawing, the water helps to cool the saw blade and also keeps the amount of saw dust down. Another type of commonly used concrete saw is called a soft-cut saw (*Figure 44*). It is designed to cut control joints in concrete within the first hour or two after finishing, rather than after the final set of the concrete. The advantage of early joint cutting is that it allows the joints to be in place

BLADE GUARD

WATER HOSE

BLADE

27305-14_F42.EPS

Figure 42 Concrete power saw.

27305-14_F43.EPS

Figure 43 Wet- and dry-cut floor-type concrete saw.

before significant tensile stresses that can cause random cracks develop in the concrete.

In addition to being used to cut control and decorative joints, power saws are also used in concrete work to cut out concrete sections and to aid in the demolition of structures.

WARNING!

Concrete dust is toxic. Respiratory protection must be worn at all times by workers using a concrete saw or working near one that is in use. It is also recommended that water be used to reduce the dust. Check your company safety practices manual.

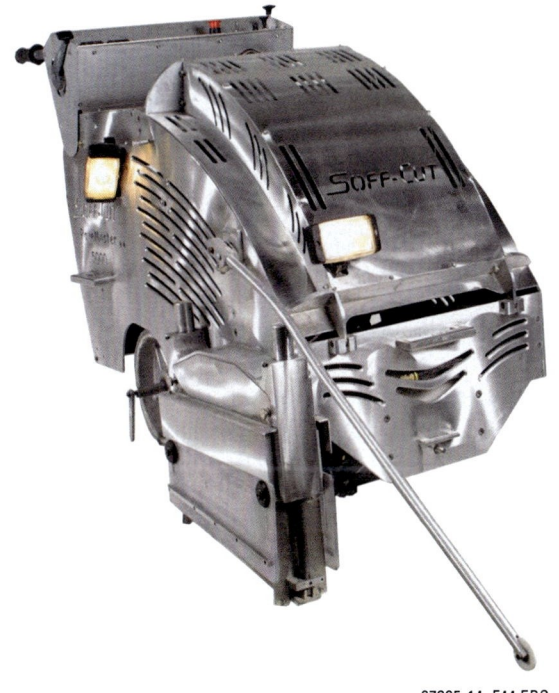

27305-14_F44.EPS

Figure 44 Soft-cut concrete saw.

5.4.0 Decorative Joints

Decorative joints (*Figure 45*) are made to create decorative patterns and designs. For example, a hand groover can be used to make a joint that divides a concrete slab into attractive spacings. Other joints can be made by placing decorative wood (redwood, cedar, or pressure-treated) inserts into the concrete to provide attractive spacings and/or designs.

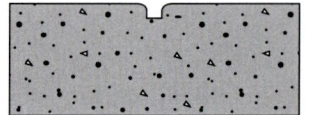
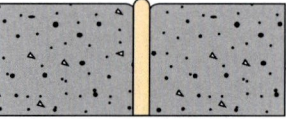

DECORATIVE DUMMY JOINT   DECORATIVE JOINT (WITH WOOD SPACER)

27305-14_F45.EPS

Figure 45 Types of decorative joints.

Identifying Construction and Control Joints on Construction Drawings

Construction joints and control joints have the same abbreviation (CJ) when shown on drawings. However, construction joints are drawn as wider lines than lines shown for control joints.

5.0.0 Section Review

1. Construction joints can be a source of _____.
   a. uncontrolled cracking
   b. layer separation
   c. water leaks
   d. settling

2. Joints installed around the perimeter of a floating floor slab are called _____.
   a. control joints
   b. isolation joints
   c. construction joints
   d. decorative joints

3. Control joints are used to prevent cracks from forming.
   a. True
   b. False

4. When creating decorative joints using wood strips, a type of wood that should *not* be used is _____.
   a. redwood
   b. cedar
   c. pressure-treated wood
   d. nontreated

SUMMARY

It is important for carpenters to understand the proper methods of handling, placing, and finishing concrete. Knowledge about the factors involved with these tasks enables the carpenter to better lay out and construct concrete forms so that the forms help, rather than hinder, efficient placement of the concrete. In addition, carpenters are often required to perform some or all parts of the concrete handling, placement, and finishing work. This is especially true if they are involved in residential, light commercial, or remodeling work. On large construction projects, carpenters normally work closely with the concrete trades; therefore, it is important for them to fully understand the concrete placement and finishing process so that their work is performed professionally and in an efficient manner.

1. Trowel-machine blades should be changed _____.

   a. when they become dull
   b. after four hours of use
   c. daily
   d. when they become ragged

2. A factor that contributes to developing cement dermatitis is _____.

   a. low air temperatures
   b. excessive sweating
   c. tight clothing
   d. contact with form oil

3. Plants that proportion dry ingredients for concrete that will be mixed in trucks are described as _____.

   a. batch plants
   b. pre-mix plants
   c. dry-mix plants
   d. blending plants

4. Trailer-mounted units that proportion ingredients and mix concrete on remote sites are known as _____.

   a. mobile mixers
   b. mobile batching plants
   c. volumetric mixers
   d. on-demand plants

5. When placing concrete, the free-fall distance should not be greater than _____.

   a. 18 inches
   b. 2 feet
   c. 3 feet
   d. 4 feet

6. Concrete placement for tall buildings is usually done by using buckets hoisted by _____.

   a. power windlasses
   b. tower cranes
   c. portable cranes
   d. cantilever booms

7. A typical concrete-pump boom truck can deliver mixed concrete to a height of _____.

   a. 250 feet
   b. 500 feet
   c. 1,000 feet
   d. 1,500 feet

8. The American Concrete Pumping Association requires that the boom be kept at least 17 feet away from _____.

   a. power lines
   b. excavation edges
   c. live steam lines
   d. personnel hoists

9. For small jobs where short distances are involved, concrete may be delivered to the point of use by concrete carts or _____.

   a. chutes
   b. wheelbarrows
   c. buckets
   d. tremies

10. On most commercial jobs, the formwork must be inspected before _____.

   a. reinforcement is installed
   b. concrete quantities are determined
   c. a permit will be issued
   d. concrete placement can begin

11. Wall forms are filled by placing concrete in layers called _____.

   a. flights
   b. lifts
   c. bands
   d. slugs

12. To prevent concrete from striking the walls in narrow forms, the concrete is placed using a _____.

   a. small concrete bucket
   b. flexible hose
   c. drop chute
   d. V-trough

13. When a long drop is necessary, concrete should be placed vertically to minimize _____.

   a. voids
   b. separation
   c. water loss
   d. stress on the formwork

14. Entrapped air is a void in concrete that has a diameter of at least _____.

   a. 1 millimeter
   b. 5 millimeters
   c. 1 centimeter
   d. 3 centimeters

15. The tool typically used for consolidating concrete in slabs, columns, beams, and walls is the _____.

   a. tamper
   b. external vibrator
   c. internal vibrator
   d. plate vibrator

16. Measurements of floor flatness to meet ASTM specifications are made with a rolling F-meter or a(n) _____.

   a. laser rangefinder
   d. profilometer
   c. electronic level
   d. dipstick floor profiler

17. A manual screed is moved across the concrete surface using _____.

   a. a sawing motion
   b. a lift-and-drop tamping motion
   c. smooth, sweeping arcs
   d. a zigzag pattern

18. On slabs that require very flat profiles, the preferred finishing tool is the _____.

   a. roller screed
   b. heavy-duty vibrating screed
   c. self-propelled laser screed
   d. magnesium manual screed

Figure 1
27305-14_RQ01.EPS

19. The concrete finishing tool shown in Review Question *Figure 1* is a(n) _____.

   a. step-and-corner tool
   b. margin trowel
   c. cove-and-base tool
   d. edger/jointer

20. For a rough decorative finish, aggregate in hardened concrete can be fractured using a _____.

   a. surfacing hammer
   b. bush hammer
   c. mason's hammer
   d. chipping hammer

21. Rough finishing of a concrete surface is done with a type of grinder known as a _____.

   a. scarifier
   b. concrete etcher
   c. power eroder
   d. surfacer

22. Joints formed at points where concrete placement is temporarily stopped are called _____.

   a. construction joints
   b. control joints
   c. expansion joints
   d. layer joints

23. To ensure a good bond between concrete layers, the surface of the previously placed concrete should be _____.

   a. as smooth as possible
   b. flooded with water
   c. slightly roughened
   d. coated with a joint sealant

24. Isolation joints can be made from _____.

   a. concrete
   b. aggregate
   c. asphalt-saturated fiber
   d. recycled steel

25. Control joints are installed to _____.

   a. compensate for concrete expansion
   b. induce cracking at desired points
   c. prevent the formation of cracks
   d. separate materials with different expansion rates

Fill in the blank with the correct term that you learned from your study of this module.

1. Eliminating voids by working freshly placed concrete to compact each layer with the one below is called _____.

2. _____ is a condition caused by the upward movement of water in concrete moving fine particles to the surface.

3. Over a fairly short period of time, concrete will harden, or _____.

4. When solids sink and water moves to the top, the concrete is said to _____.

5. A(n) _____ material displays excellent flexibility and elongation properties.

6. An improper concrete placement may result in _____, a separation of the sand and cement from the gravel.

7. The generic term for concrete breakup (crumbling, splitting, or chipping) is _____.

8. A vertical divider inserted in a form to prevent further concrete spread is called a(n) _____.

9. A uniform layer of concrete is described as a(n) _____.

10. Mixing concrete in a batch plant until a slump test indicates the desired slump can be achieved, then completing the mixing in the delivery truck is described as _____.

11. A(n) _____ connection is hinged or pivoted.

12. _____ is a thin mortar.

Trade Terms

Articulating	Elastomeric	Segregation
Bleed	Grout	Set
Bulkhead	Laitance	Shrink mixing
Consolidating	Lift	Spalling

Trade Terms Introduced in This Module

Articulating: Having a hinged or pivoting connection.

Bleed: A condition in concrete in which the solids settle and the water moves to the top.

Bulkhead: A vertical piece placed inside a concrete form to stop concrete at a certain point. A construction joint is created at the bulkhead.

Consolidating: Working freshly placed concrete so that each layer is compacted with the layer below and voids caused by water or air pockets are eliminated.

Elastomeric: A material having the properties of excellent flexibility and elongation.

Grout: A thin mortar.

Laitance: Fine particles on a concrete surface resulting from an upward movement of water in the concrete.

Lift: Concrete layer of uniform thickness.

Segregation: The separation of sand-cement ingredients from the gravel due to the improper placement of concrete.

Set: The hardening of concrete.

Shrink mixing: Practice of mixing concrete at a batch plant to the point where the slump meter indicates that the desired slump is predictable, then finishing mixing the concrete on the way to the job site.

Spalling: The condition of concrete breakup (chipping, splitting, or crumbling).

Additional Resources

This module presents thorough resources for task training. The following resource material is suggested for further study.

American Concrete Institute (ACI). **www.concrete.org**

American Concrete Pumping Association. **www.concretepumpers.com**

ASTM E1155, Standard Test Method for Determining F_F Floor Flatness and F_L Floor Levelness Numbers, latest edition. West Conshohocken, PA: ASTM International.

Cement Association of Canada. **www.cement.ca**

Portland Cement Association. **www.cement.org**

Figure Credits

Section Review Answer Key

Answer	Section Reference	Objective Reference
Section One		
1. b	1.1.0	1a
2. a	1.2.0	1b
3. c	1.3.0	1c
Section Two		
1. d	2.0.0	2
2. c	2.1.0	2a
3. a	2.2.7	2b
Section Three		
1. b	3.1.1	3a
2. c	3.2.0	3b
Section Four		
1. c	4.0.0	4
2. d	4.1.0	4a
3. c	4.2.0	4b
4. b	4.3.4	4c
5. a	4.4.0	4d
6. c	4.5.0	4e
7. False	4.4.0	4d
Section Five		
1. c	5.1.0	5a
2. b	5.2.0	5b
3. False	5.3.0	5c
4. d	5.4.0	5d

NCCER – *Carpentry Level Three*

NCCER CURRICULA — USER UPDATE

NCCER makes every effort to keep its textbooks up-to-date and free of technical errors. We appreciate your help in this process. If you find an error, a typographical mistake, or an inaccuracy in NCCER's curricula, please fill out this form (or a photocopy), or complete the online form at **www.nccer.org/olf**. Be sure to include the exact module ID number, page number, a detailed description, and your recommended correction. Your input will be brought to the attention of the Authoring Team. Thank you for your assistance.

Instructors – If you have an idea for improving this textbook, or have found that additional materials were necessary to teach this module effectively, please let us know so that we may present your suggestions to the Authoring Team.

NCCER Product Development and Revision

13614 Progress Blvd., Alachua, FL 32615

Email: curriculum@nccer.org
Online: www.nccer.org/olf

❏ Trainee Guide ❏ Lesson Plans ❏ Exam ❏ PowerPoints Other _____

Craft / Level: _____ Copyright Date: _____

Module ID Number / Title: _____

Section Number(s): _____

Description: _____

_____

_____

_____

Recommended Correction: _____

_____

_____

_____

Your Name: _____

Address: _____

_____

Email: _____ Phone: _____